Gary E. McCuen

IDEAS IN CONFLICT SERIES

publications inc.

502 Second Street
Hudson, Wisconsin 54016

All rights reserved. No part of this book may be reproduced or stored by any means without prior permission from the copyright owner.

Illustration & Photo Credits

Sack 11, 87, 118, 143, Stayskal 15, Natural Resources Defense Council 21, 28, 133, 150, Carol*Simpson 37, 61, 122, 129, The United Farm Workers 49, American Council on Science and Health 55, 70, 137, Locher 66, 76, International Atomic Energy Agency 91, National Coalition to Stop Food Irradiation 99. Cover illustration by Ron Swanson.

©1991 by Gary E. McCuen Publications, Inc.
502 Second Street, Hudson, Wisconsin 54016
(715) 386-7113

International Standard Book Number 0-86596-078-X Printed in the United States of America

CONTENTS

Ideas in Conflict   6

CHAPTER 1   IS OUR FOOD SUPPLY SAFE?

1. TOO MANY POISONS IN OUR FOOD   9
   Al Meyerhoff

2. AN EPIDEMIC OF FEAR   13
   MIchael Specter

3. HIDDEN HAZARDS IN MEAT AND POULTRY   19
   Mary Deery Uva

4. THE SAFEST MEAT SUPPLY IN THE WORLD   26
   C. Manly Molpus

CHAPTER 2   CHEMICALS AND FOOD PRODUCTION

5. THE HEALTH RISKS OF MAN-MADE PESTICIDES   35
   Ralph Nader

6. THE HEALTH RISKS OF NATURE-MADE PESTICIDES   41
   William R. Havender

7. THE DANGERS OF POISON CHEMICALS   47
   Cezar Chavez

8. THE BENEFITS OUTWEIGH THE RISKS   53
   Leonard T. Flynn

9. INTOLERABLE RISK: PESTICIDES IN CHILDREN'S FOOD   59
   Mothers and Others

10. LITTLE PESTICIDES FOUND IN CHILDREN'S FOOD   64
    Food and Drug Administration

11. THE DIOXIN COVER-UP   68
    Peter Von Stackelberg

12. DIOXIN LEVELS ARE NOT HAZARDOUS   74
    Red Cavaney

13. THE DANGERS OF IMPORTED FOODS: 80
POINTS AND COUNTERPOINTS
Thomas M. Dorney vs. William L. Schwerner

CHAPTER 3 IRRADIATING FOODS

14. THE BENEFITS OF FOOD IRRADIATION 89
American Council on Science and Health

15. THE HAZARDS OF FOOD IRRADIATION 95
National Coalition to Stop Food Irradiation

CHAPTER 4 SEAFOOD SAFETY

16. CONSUMING SEAFOOD IS RISKY BUSINESS 106
Ellen Haas

17. OUR SEAFOOD IS SAFE TO EAT 112
Lee J. Weddig

CHAPTER 5 ORGANIC FARMING AND BIOTECHNOLOGY

18. ORGANIC FOODS: AN INTRODUCTION 120
Ginia Bellafante

19. THE CASE FOR ORGANIC FOODS 127
Colman McCarthy

20. THE CASE AGAINST ORGANIC FOODS 131
Warren T. Brookes

21. BOTH ORGANIC FARMING AND PESTICIDES 135
ARE NEEDED
John Hood

22. BIOTECHNOLOGY AND AGRICULTURE: 141
THE POINT
W. P. Norton

23. BIOTECHNOLOGY AND AGRICULTURE: 148
THE COUNTERPOINT
John Nicholson

Bibliography 156

REASONING SKILL DEVELOPMENT?

These activities may be used as individualized study guides for students in libraries and resource centers or as discussion catalysts in small group and classroom discussions.

1. Examining Counterpoints: Toxic Pesticides and Scientific Evidence — 31
2. Interpreting Editorial Cartoons — 86
3. Points and Counterpoints: Natural Pesticides vs. Man-Made Pesticides — 102
4. Interpreting Editorial Cartoons — 118
5. What is Editorial Bias? — 154

This series features ideas in conflict on political, social, and moral issues. It presents counterpoints, debates, opinions, commentary, and analysis for use in libraries and classrooms. Each title in the series uses one or more of the following basic elements:

Introductions *that present an issue overview giving historic background and/or a description of the controversy.*

Counterpoints *and debates carefully chosen from publications, books, and position papers on the political right and left to help librarians and teachers respond to requests that treatment of public issues be fair and balanced.*

Symposiums *and forums that go beyond debates that can polarize and oversimplify. These present commentary from across the political spectrum that reflect how complex issues attract many shades of opinion.*

A **global** *emphasis with foreign perspectives and surveys on various moral questions and political issues that will help readers to place subject matter in a less culture-bound and ethnocentric frame of reference. In an ever-shrinking and interdependent world, understanding and cooperation are essential. Many issues are global in nature and can be effectively dealt with only by common efforts and international understanding.*

Reasoning skill *study guides and discussion activities provide ready-made tools for helping with critical reading and evaluation of content. The guides and activities deal with one or more of the following:*

RECOGNIZING AUTHOR'S POINT OF VIEW

INTERPRETING EDITORIAL CARTOONS

VALUES IN CONFLICT

WHAT IS EDITORIAL BIAS?

WHAT IS SEX BIAS?

WHAT IS POLITICAL BIAS?

WHAT IS ETHNOCENTRIC BIAS?

WHAT IS RACE BIAS?

WHAT IS RELIGIOUS BIAS?

*From across **the political spectrum** varied sources are presented for research projects and classroom discussions. Diverse opinions in the series come from magazines, newspapers, syndicated columnists, books, political speeches, foreign nations, and position papers by corporations and nonprofit institutions.*

About the Editor

Gary E. McCuen is an editor and publisher of anthologies for public libraries and curriculum materials for schools. Over the past years his publications have specialized in social, moral and political conflict. They include books, pamphlets, cassettes, tabloids, filmstrips and simulation games, many of them designed from his curriculums during 11 years of teaching junior and senior high school social studies. At present he is the editor and publisher of the *Ideas in Conflict* series and the *Editorial Forum* series.

Chapter 1

IS OUR FOOD SUPPLY SAFE?

1. TOO MANY POISONS IN OUR FOOD 9
   Al Meyerhoff

2. AN EPIDEMIC OF FEAR 13
   Michael Specter

3. HIDDEN HAZARDS IN MEAT AND POULTRY 19
   Mary Deery Uva

4. THE SAFEST MEAT SUPPLY IN THE WORLD 26
   C. Manly Molpus

READING

1   IS OUR FOOD SUPPLY SAFE?

TOO MANY POISONS IN OUR FOOD

Al Meyerhoff

Al Meyerhoff, a senior attorney with the Natural Resources Defense Council, wrote this article for the Los Angeles Times.

Points to Consider:

1. How extensive is chemical use each year by American farmers?

2. Why is the Environmental Protection Agency (EPA) criticized by the author?

3. What kind of legislation should the U.S. Congress pass?

4. What is our best hope for a safe food supply?

Al Meyerhoff "Poison In The Food We Eat" *Star Tribune,* May 18, 1989. Reprinted with permission of Al Meyerhoff, Natural Resources Defense Council.

It is children who are squarely at the highest risk from an abysmal record of government inaction.

Just what are we doing to our food? Each year in producing the American food supply, farmers use up to two billion pounds of chemicals to kill insects, eliminate fungus, destroy weeds, "control" plant growth, increase yields and promote uniform artificial ripening. Many of these chemicals are the most toxic of all human creations, originating from chemical-warfare research during World War II and exploding into use during the biochemical revolution of the 1950s and 1960s.

Toxins

Where do these toxins go? Nobody knows. The chemical companies have never adequately tested most of them to identify their long-term chronic health effects on humans—effects such as cancer, genetic mutations, birth defects or neurological disorders. And the methods used by the Food and Drug Administration to test food for "illegal" residues of these poisons can detect less than half of the chemicals used on food.

Lately, national attention has been focused on apples, and a pesticide called Alar and its breakdown product, known as UDMH, a powerful carcinogen. Actually, the problem isn't just apples or Alar—it's a failed government regulatory system and an unwillingness by agriculture to accept changes in its use of toxic chemicals.

Necessary Evil

Each of us gets our daily dose of toxic chemicals in the food we eat. Three hundred different pesticides are used in our food, leaving residues in fruit, vegetables, grain products, even in milk, eggs, poultry and meat. We have been told that this represents a "necessary evil", the price of our abundant, inexpensive food supply. But is it? And at what cost?

Last year, the Environmental Protection Agency decided that the health risks presented by pesticides in food were its number one priority. "Pesticides dwarf the other environmental risks the agency deals with," said a high-level EPA official. "Toxic waste dumps may affect a few thousand people who live around them. But virtually everyone is exposed to pesticides."

Yet since then the agency has continued to act feebly or not at all to protect the food chain. For example, last month it was disclosed that EPA inaction on an extraordinary toxic pesticide called Aldicarb has resulted in up to 80,000 children on any given day being exposed to dangerous levels of this toxic substance in potatoes alone.

Reprinted by permission of *Star Tribune*.

Children

It is children who are squarely at the highest risk from an abysmal record of government inaction. The results of a two-year study released by scientists from the Natural Resources Defense Council found that:

- EPA's supposedly "safe" pesticide residue levels, set up to 20 years ago, were based on the assumed diet of a fully grown adult male and ignored the far different eating habits of children. The cancer risks to children from the actual residues of the eight pesticides analyzed by NRDC were up to several hundred times higher than the cancer risk level designated "acceptable" by EPA.

- Between 5,500 and 6,200 preschoolers might get cancer solely as a result of their exposure by age six to the eight common carcinogenic pesticides in fruits and vegetables analyzed in the study. (The EPA has classified 66 pesticides as potential human carcinogens.)

- Fifty-five percent of the lifetime risk of developing cancer from exposure to carcinogenic pesticides used on fruit is typically incurred by the time a child reaches age six.

> ### DRUGS IN MILK
>
> *The dairy industry and government inspectors routinely fail to detect traces of a number of potentially dangerous drugs in milk supplies, according to a report Friday in the Wall Street Journal.*
>
> *Two recent surveys showed that some low-fat and skim milk is contaminated with antibiotics and other drugs, including one suspected carcinogen, that are used to treat sick cattle. The industry and its federal regulators fail to inspect for those drugs.*
>
> "Tests Often Miss Drugs In Milk," Star Tribune, December 30, 1989, p. 7A

Legislation

What's to be done? Congress must enact comprehensive legislation, recently introduced by Rep. Henry Waxman, D-Calif., and Sen. Edward Kennedy, D-Mass., setting strict health-based limits on pesticide residues, protecting especially vulnerable subpopulations, such as children, and streamlining procedures to get the worst chemicals off the market quickly. And a well-known group of old, potent carcinogens presenting the bulk of the cancer risk to the American public should be phased out over the next few years. Chemical reduction goals as a condition of multibillion-dollar price supports for surplus crops is an idea worth exploring in next year's farm bill debate.

Best Hope

But perhaps our best hope for a safe food supply will come from consumer pressure. Major supermarket chains in California, Colorado and Massachusetts—and in Great Britain—are already offering consumers a choice between chemical-free and chemically treated produce.

Individual consumers must now act on their concerns and choose health. Change is coming, and let us hope that it is with the active participation of the American farmer.

READING 2
IS OUR FOOD SUPPLY SAFE?

AN EPIDEMIC OF FEAR

Michael Specter

Michael Specter wrote the following article for the Washington Post.

Points to Consider:

1. What kind of risks do people worry about most?

2. How are the most serious threats to the nation's health described?

3. Why are people healthier than ever before?

4. How do people make judgments about risks in their lives?

5. Explain the term "acceptable risk".

© 1989, Washington Post Writers Group. Reprinted with permission.

Battered by an almost daily torrent of worrisome reports about the hazards of what they eat and how they live, Americans have become engulfed in an epidemic—not of cancer but of fear.

Like so many of her friends and neighbors, Ellen Corella has replaced well-marbled meats and cream sauces in her diet with water-packed tuna and vegetable puree.

She reads food labels, pays attention to sodium and pushes fiber at her family every chance she gets. Cruising the grocery shelves in pursuit of organic jewel yams and macrobiotic sweetener, the Maryland housewife has transformed herself into a paragon of defensive shopping.

Risk and Fear

"I'm just doing what I can to cut down our risks," she said recently, while negotiating an aisle full of bottled waters at a natural food shop in the district. "When you think about it, these are life-and-death matters. I don't want my kids eating a bunch of chemicals, and I don't know too many people who do."

Battered by an almost daily torrent of worrisome reports about the hazards of what they eat and how they live, Americans have become engulfed in an epidemic—not of cancer but of fear. Increasingly, people see grave risks in the most basic elements of their lives: their food, their water, even the air they breathe.

Obsessive Reliance

Eating too much fat has clearly proved to be a risk worth worrying about. But many scientists say an obsessive reliance on bottled water and organic produce is foolish. They say these fears make little sense and that Americans too often overreact to the most trivial of risks while ignoring much more substantial threats to their health or safety. Over and again, they note that three things cause the vast majority of premature deaths in the United States: alcohol, tobacco, and eating too much saturated fat.

"It worries me greatly, but the facts don't seem to help much," said Surgeon General C. Everett Koop. "People just have an inappropriate sense of what is dangerous. They get overly upset about minor problems. If you translate the weight of a laboratory rat and time it takes him to develop bladder cancer to a 200 pound man drinking Fresca (which contains artificial sweeteners), it comes out to about two bathtubs full each day. People dropped Fresca in a minute, but they continue to smoke."

Reprinted by permission: *Tribune Media Services.*

Never Healthier

The truth is that Americans have never been healthier. Average life expectancy has risen steadily for decades and, except for cancers caused by smoking and exposure to the sun, cancer death rates have dropped or remained relatively stable. There is no cancer epidemic except for the lung cancer, which is traced almost entirely to smoking. Yet surveys have repeatedly shown that people have never been more anxious about their health.

"People just seem to see the apocalypse everywhere they turn," said Bruce Ames, chairman of the department of biochemistry at the University of California at Berkeley, who was among the first to point out that natural pesticides are at least 10,000 times more common than those made by man. "There are some important risks, of course. But everyone should just relax a bit and have some fun."

At times that seems hard to do. Provocative warnings about too much cholesterol, not enough vitamin A and what can happen to people who do not exercise enough—or do not exercise properly—have become part of the tapestry of American life. To some, cancer seems hidden in every meal.

PEOPLE ARE HEALTHIER

These days news seems to be synonymous with bad news. We hear our environment is contaminated with pesticides, additives, dioxin and more. We shiver when we face the latest "carcinogen of the week" and yearn for the good old days when life was more "natural" and people were healthier.

The truth, however, is that the American people are healthier, living longer, and have more control over their personal well-being than ever before in history. Our technological way of life has contributed to good health, not detracted from it. Life expectancy has reached new highs; infant mortality has plummeted; the overall death rate has declined. Even heart disease mortality—the leading cause of death—and cancer mortality are down.

Media Update, American Council on Science and Health, January-December, 1988, p.1

Risks and Accidents

Yet, any person in the United States is thousands of times more likely to die of a household accident or car wreck than of cancer caused by man-made pesticides in food. But in interviews with 25 shoppers buying organic produce at a local supermarket last week, nearly half said they had not worn their seat belts on the way to the store.

"Driving is a pretty risky proposition when you compare it to, say, drinking apple juice with a trace amount of Alar in it," said Richard Wilson, professor of physics at Harvard University, an expert in comparing risks. "But everyone thinks he's a better driver than the next guy. Alar is something we have no control over—we can't even be sure if it's in there or not."

"It's hard for most people to develop a perspective on risks," he continued. "After all, how many of us really know what it means when they say the risk of something is one in a million?"

Sorting Out Risks

Sorting out the risks in a normal life is tricky business.

For example, more than 30 percent of regular smokers will die from some disease connected to their habit, losing an average of 8.3 years from normal life expectancy. But many people react equally or more forcefully to the evidence that there may be a one in a million risk of getting cancer from chemicals found in drinking water.

One reason most risks are difficult to gauge is that most people do not reason in purely statistical terms about their lives. Also, most people make deep distinctions between risks they can control, such as smoking, and those foisted on them, such as asbestos in the walls of a school.

"It doesn't really do much good to say cigarette smoking kills far more people than pesticides, so let's forget about pesticides," said Michael Jacobson, executive director of the Center for Science in the Public Interest. "The fact is that these things, like Alar, appear to be a modest problem and we shouldn't ignore them simply because other problems are worse."

Science and Certainty

In many cases, people appear to be searching for a degree of certainty that science cannot provide.

How directly will weight loss affect the development of heart disease? Is a sunburn really going to cause cancer? If a rat gets cancer from consuming a huge amount of diet soda in a short time, will two sodas a day for 30 years do the same thing to a human? Why did Uncle Jake live to be 90 if eating steak every day is so bad?

None of those questions can be answered with finality. The link between diet and heart disease has been made firmly over the past decade. But genetics often plays a major role that confounds the dietary odds. And researchers have long known that the easiest way to reduce the cancer rate in laboratory animals is simply to give them less food.

The incidence of skin cancer caused by exposure to the sun is increasing rapidly, but nearly everybody knows healthy elderly people who have spent their lives in the sun without using a drop of sun block. Obesity contributes dramatically to death from heart disease, according to dozens of government studies, but one need not look long before finding old people who are fat.

"That's what I call the 'my uncle had an Edsel and it ran fine' problem," said Park Dietz, a professor of law and psychiatry at the University of Virginia, and a specialist on the behavior of large groups. "In most people's thinking, a single familiar case outweighs all the data from the opposite direction."

Acceptable Risk

People have a sliding scale of what they consider to be an acceptable risk. Unknown fears always seem to outrank familiar ones. Driving, drinking or combining the two is something a person chooses to do. Consuming fruit with pesticides in it usually is not.

Equal risks are rarely treated equally. Aflatoxin and dioxin are both among the most potent of cancer-causing chemicals. But dioxin, which is manufactured in tiny amounts during processes such as bleaching paper, is often opposed by environmental groups because it is an artificial chemical.

Nobody has tried to ban aflatoxin, a natural carcinogen that can frequently be found in similarly minuscule amounts in such common foods as peanut butter.

"Nobody knows anybody who got cancer from eating a peanut butter sandwich," said Harvard's Wilson. "But the risk of eating a peanut butter sandwich every day is greater than the dioxin risk if it's calculated in a similar way."

Most people also have a tough time deciding what is an acceptable risk. That too depends on the occasion. Skiing may be risky, but it is also fun. If a child's toy posed the same level of risk, few people would buy it.

Radiation Fears

Radiation fears rank among the best examples of the sliding, and often irrational, scale many people apply to risks in their lives.

It is not enough, for example, to suggest that people fear radiation, though in certain instances they obviously do.

Medical X-rays scare very few people even though they can cause cancer, and public health officials have tried for years to persuade Americans to question their physicians more aggressively. But nuclear energy is opposed with compelling vigor by many Americans, and even those who support it consider it dangerous.

Radon, an odorless gas that can seep into homes from the ground, is radioactive and is by far the greatest source of radiation exposure to the average person in the United States. Despite recent publicity, relatively few people have had their homes tested for its presence.

"Coal is more dangerous in many ways than any radiation risk," said Paul Slovic, president of Decision Research in Eugene, Oregon, and professor of psychology at the University of Oregon. Miners are killed every year to obtain coal. And burning it releases more radioactivity than has escaped from nuclear power plants. "But coal is nothing new. It comes from the ground, and it's natural. You just set fire to it, and you are warm and comfortable. People welcome what they understand and fear what they don't. It's just a very hard problem to solve."

READING
3
IS OUR FOOD SUPPLY SAFE?

HIDDEN HAZARDS IN MEAT AND POULTRY

Mary Deery Uva

This pamphlet was prepared by Mary Uva, Science Associate at the Natural Resources Defense Council. The Natural Resources Defense Council is a nonprofit membership organization dedicated to protecting natural resources and to improving the quality of the human environment. NRDC's programs of legal action, scientific research, and citizen education are supported by 70,000 members nationwide.

Points to Consider:

1. Identify the two most common disease-causing bacteria found in meat and poultry.

2. How many people get sick each year from bacteria in meat?

3. What antibiotics are given to treat people who are fighting infection caused by bacteria?

4. How and why are these antibiotics used in the meat industry?

Mary Deery Uva, "Hidden Hazards in Meat and Poultry", Natural Resources Defense Council, 1989.

New scientific evidence shows that these two problems, the increasing number of foodborne disease outbreaks and the declining power of antibiotics in health care, are clearly linked at the barnyard feed trough.

Every year millions of Americans get sick and thousands die as a result of food poisoning caused by bacterial contamination, and the number of victims is increasing every year. Many of these cases occur because our meat and poultry supplies are extensively contaminated with disease-causing bacteria such as *Salmonella* and *Campylobacter*.

In the past, serious cases of bacterial food poisoning—often marked by agonizing symptoms such as nausea, vomiting, diarrhea, cramps, headaches and fever—have been successfully treated with antibiotics such as penicillin and tetracycline. In recent years, however, doctors have found that antibiotics are just not working as well as they once did.

New scientific evidence shows that these two problems, the increasing number of foodborne disease outbreaks and the declining power of antibiotics in health care, are clearly linked at the barnyard feed trough. Cattle, hogs, and poultry are routinely fed antibiotics to promote meat production, a practice that is compromising the effectiveness of these drugs in human health care.

NRDC has prepared this guide to explain how these hidden hazards in our meat have come to be there, what you can do to avoid becoming ill, and how you can help press for better public protection from these hazards.

Hidden Hazards in Meat and Poultry

Salmonella, Campylobacter, and other disease-causing bacteria are common contaminants in fresh meat and poultry. These bacteria usually originate in animals raised in factory farms and, during slaughtering and processing, spread easily from infected animal tissues to meat processing equipment and raw meat and poultry products. As a result, bacterial contamination is a widespread problem in the U.S. meat and poultry supply. *Salmonella* contamination alone is found in almost 40 percent of all poultry, 12 percent of pork, and 5 percent of beef.

Consumers run the risk of exposure to these disease-causing bacteria through improper handling or cooking of meat and poultry. Some increase the risks further by unknowingly cross-contaminating other foods with unwashed utensils that were used to prepare raw meat or poultry.

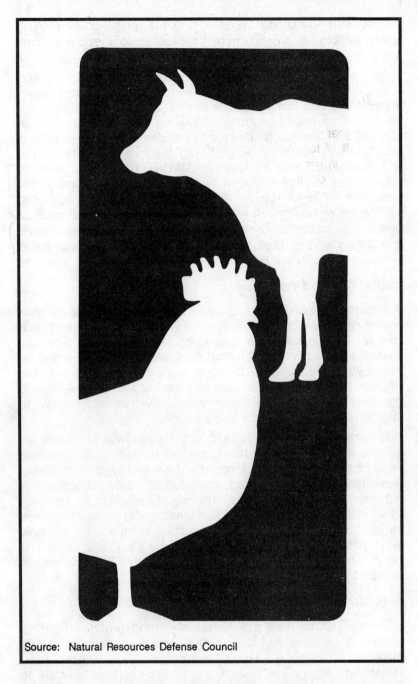

Source: Natural Resources Defense Council

Most normal healthy adults who get sick from foodborne bacteria experience mild symptoms that go away in a few days without any medical treatment, and many never realize the cause of their illness. But in a small number of cases, foodborne infections can be very serious and require antibiotic therapy and/or hospitalization. Young children, the elderly, people with weakened immune systems, and people already taking antibiotics are far more susceptible to these infections and more likely to require medical treatment.

Approximately 2 million people get *Salmonella* poisoning and 2.1 million more get *Campylobacter* poisoning every year. Medical bills and lost work time attributed to these infections are estimated to cost consumers between $1.3 and $2.4 billion annually. Unfortunately, despite the epidemic number of victims and the high costs associated with foodborne infections, the U.S. Department of Agriculture (USDA), the federal agency responsible for the safety of our meat supply, has no safety standards for bacteria such as *Salmonella* or *Campylobacter* in fresh meat or poultry. The USDA does not inspect raw meat or poultry for bacterial contamination and instead believes that consumers should bear the responsibility for controlling these hidden hazards in our meat.

Antibiotics and You

Antibiotics—penicillin, tetracycline, streptomycin, lincomycin and erythromycin to name a few—are vitally important in health care for fighting infections caused by bacteria. When you take an antibiotic to treat an infection, it works by weakening or destroying the bacteria in your body that are causing the infection. Although the antibiotic is aimed at the disease-causing bacteria, most other bacteria in your body that are sensitive to that particular antibiotic are also destroyed.

Unfortunately, antibiotics don't always work. When an antibiotic is overused, the exposed bacteria can become resistant to it, and can no longer be destroyed by that antibiotic. Some resistant infections can be treated with different antibiotics, but many bacteria become simultaneously resistant to as many as six different antibiotics, seriously limiting the available choices for effective treatment. Antibiotic-resistant infections are not only more difficult and expensive to treat, but they also often last longer and require hospitalization more frequently than infections caused by antibiotic-sensitive bacteria.

Animals + Antibiotics = Super-Bugs on Meat

Though antibiotics are extremely important for fighting infectious diseases in humans, they are also used for a very different purpose—many livestock producers feed sub-therapeutic or low doses of antibiotics to food animals to make them grow faster and produce more meat. Some of these feed additives are the same antibiotics used to treat human infections.

As a result of widespread use of antibiotics by the livestock industry, many bacteria in animals—and in raw meat and poultry—are antibiotic-resistant. Thus, foodborne infections caused by bacteria such as *Salmonella* or *Campylobacter* are frequently resistant to the same antibiotics that we normally take to fight them. For example, ampicillin, a semisynthetic form of penicillin, has always been one of the drugs of choice for treating severe *Salmonella* infections, but now many *Salmonella* strains are ampicillin-resistant and cannot be controlled with this drug.

The resistance generated by antibiotic use in livestock is not limited to foodborne bacteria. Antibiotic-resistant bacteria can spread their resistance to other bacteria, making them antibiotic-resistant. If you are exposed to antibiotic-resistant bacteria on meat or poultry, the antibiotic resistance can spread to other potentially disease-causing bacteria which may be present in your body.

Antibiotic Resistance and Your Health

Physicians often overlook resistance problems and prescribe antibiotics without first determining whether an antibiotic will be effective. If the first drug does not work, you might have to try a number of different antibiotics before finding one that does work. Improper or ineffective antibiotic therapy can in fact make some antibiotic-resistant infections worse by destroying normal bacteria in your body that would otherwise help keep the infection in check. This often prolongs the infection and costs you much more for expensive alternative antibiotics and extended medical care.

Sometimes antibiotic therapy for one infection can actually induce a second, antibiotic-resistant infection. In antibiotic-resistant *Salmonella* outbreaks, many people get sick because they are already taking antibiotics for other infections when they are exposed to the *Salmonella* bacteria. While the antibiotics destroy most of the normal bacteria in your body, the antibiotic-resistant *Salmonella* survive and grow rapidly to replace the other bacteria. Under these conditions, they grow quickly to high levels, and can turn a mild or even unnoticed case into a very serious infection.

Buying Safer Meat and Poultry

Some beef, chicken, and turkey producers have reduced or eliminated the use of antibiotics (and many other chemical and hormone additives) in their animal feeds. Their products are usually well-labelled and available in many stores, from small specialty food shops to some major supermarket chains. If you can't find them in your local food store, you should ask your store manager to carry meat and poultry that is raised without antibiotics. There is a catch: these products are usually more expensive than other meat and poultry items.

Natural" Really Natural?

 current USDA regulations, meat and poultry producers
 use antibiotics and certain other feed additives and still call
 product "natural". Therefore, unless the label specifically
 states that no antibiotics were used for growth promotion,
"USDA–certified natural" meat or poultry may come from
animals raised on antibiotics. Meat and poultry may also have
residues of drugs and pesticides unless the label states, for
example, that no hormones were used or that there were no
pesticides in the animal feed. Watch out for potentially
misleading terms such as "natural feeding program" (USDA says
there's no such thing) or "no feed preservatives" (antibiotics are
not preservatives).

Supporting New Food Safety Laws

Congress, the Food and Drug Administration (FDA), and the
U.S. Department of Agriculture (USDA) have become concerned
about bacterial contamination and antibiotic resistance in our
food supply, and are now considering action to improve food
safety. Your letters of support will encourage them to act
quickly to develop safer meat production practices and remedy
these problems in our food supply.

- The FDA is responsible for regulating the use of antibiotics in livestock production. Write the Commissioner of FDA, and urge him to prohibit the use of clinically important antibiotics in animal feeds in order to reduce the public health threat from antibiotic-resistant diseases transmitted through livestock products.

- The USDA is responsible for developing ways to control the incidence of bacteria such as *Salmonella* and *Campylobacter* on raw meat and poultry. Write the Administrator of the Food Safety and Inspection Service at USDA, and urge him to develop mandatory standards and control programs for disease-causing bacteria in raw meat and poultry. Ask USDA to require all raw meat and poultry to be labelled with cooking and handling instructions to prevent food poisoning.

- Write to your Senators and Representatives and ask them to support legislation that will require the meat industry to reduce bacterial contamination in food. Let them know you are also concerned about antibiotic resistance in health care caused by the misuse of antibiotics in livestock production. Ask them to move quickly on legislation to prohibit the use of medically important antibiotics in animal feeds.

- Ask the FDA, USDA, and your Senators and Representative to develop a standardized labelling program so that consumers can easily and confidently recognize "natural" or "chemical-free" meat and poultry in stores.

READING 4

IS OUR FOOD SUPPLY SAFE?

THE SAFEST MEAT SUPPLY IN THE WORLD

C. Manly Molpus

C. Manly Molpus is the President and Chief Executive Officer of the American Meat Institute

Points to Consider:

1. Why are meat products safe to eat?

2. What problems can arise from improper cooking of meat products?

3. How serious is foodborne illness?

4. Why has meat inspection improved?

5. What statistics are given about the safety of the food supply?

Excerpted from testimony by C. Manly Molpus before the House Committee on Government Operations, April 11, 1989.

Given the safety record of meat products, we fail to see how anyone can contend that the industry does not produce the highest quality products possible.

In a recent hearing in the Senate, several witnesses attempted to create the serious misperception that this country's meat and poultry inspection system is in disarray, and that the public is at risk. Nothing could be further from the truth than the misperceptions created by those statements. While there has been disagreement on proposed regulations to implement improved processing inspection, there can be no questioning the fact that our country has the world's safest meat supply, and the issue is how, with advances in science and technology, can we make it even safer.

To counter any more spurious claims which might be made before your Subcommittee, I would like to set the record straight. For the record:

Meat Products Are Safe

While it is true that naturally-occurring bacteria may be found on raw meat, proper cooking makes these products safe. Over all of the years that the Center for Disease Control has maintained records on foodborne illness, 96% of the mishandling of meat and poultry which causes illness occurs in the kitchen. And yet, even with this potential for mishandling, meat was implicated in only 32 outbreaks of foodborne illness for the latest year reported. With a population of almost 250 million, most of which consume meat products every day, this is a safety record of which the industry can be proud.

Reported positive cultures for *Salmonella* in the U.S. have risen, but this is primarily due to improved reporting systems and better methodology among health professionals. In fact, the only year that such reports exceeded the 40,000 cited at the hearing was in 1985, when 16,661 cases were reported from one outbreak at a dairy. Additionally, the rate of increase in reported *Salmonella* has been reduced to less than half of what it was 25 years ago. Again, meat was only implicated in 16% of the Salmonellosis outbreaks over the last 12 years CDC has records for—a far cry from the "more than half" cited by one witness. Finally some perspective is necessary—based on their own data, and if they had a population as large as the U.S., the U.K. would have 60,000 reported cases of Salmonellosis each year, Sweden would have 88,000 and our neighbor Canada would have 116,000.

Foodborne Illness

Foodborne illness is serious, and public health protection is a major responsibility. However, it is not helped by misstating

Source: Natural Resources Defense Council

statistics to unnecessarily alarm the public. For the latest year that CDC has compiled data, there were just over 650 reported outbreaks of foodborne illness from all sources, a number similar to the rate for other recent years. Of the 220 which could be associated with a specific food, meat was implicated in only 14.5%, very close to the number from fruit and vegetable dishes (12.3%), and less than seafood (18.2%) which is consumed in much smaller quantities. As discussed above, what is important to recognize is that any food can lead to illness if improperly handled, cooked or stored, and no single food is the cause of foodborne illness.

Meat Inspection Has Been Improved

In 1985 the National Academy of Sciences issued a report calling for the USDA to move toward a risk-based inspection system. USDA is doing this by requiring the industry to perform more of the cutting and trimming tasks which were once performed by USDA inspectors. The industry is also required to perform more Quality Control audits, based on modern procedures. USDA inspectors monitor each of these programs, and still individually inspect each and every meat carcass in this country, 119 million meat carcasses last fiscal year. Improved Processing Inspection, which was much discussed at the hearing, would only apply to meat products being made from

> **OUTRAGEOUS REPORTS**
>
> *Quite frankly, I become outraged when I read testimony before Congressional Committees and reports in the newspapers that are inspired by those who portray the meat and poultry industry in a derogatory manner, citing disgusting conditions and unsafe handling practices. I would like to assure you that, in my almost thirty years experience in working for this (and a predecessor) trade association, and personal visits to hundreds (maybe thousands!) of meat and poultry plants of all kinds, including slaughter and processing operations, I have never personally witnessed the kinds of conditions that the Government Accountability Project has testified about in the past.*
>
> Western States Meat Association, Congressional Testimony, April 6, 1989

carcasses which had already been "Inspected and Passed" by USDA inspectors.

The only change is that USDA inspectors are doing fewer tasks, freeing up resources to increase testing for chemical and microbiological residues, neither of which was tested for under the old-fashioned system. In fact, all of the discussion of foodborne illness caused by microbiological or chemical residues is totally irrelevant to the Traditional Inspection System, since line inspectors cannot now, and have never been able to, inspect for these residues. Thus, it is counter to the facts to contend that the inspection system is being diminished in any way, when in fact it is being strengthened, and proposed changes will further modernize public health protection.

The Meat Industry Produces High-Quality, Wholesome Products

USDA inspectors have the power to shut down any meat processing facility which fails to meet standards, and have not hesitated to do so in the past. In addition, the meat industry is completely dependent upon repeat business, with every manufacturer putting its reputation on the line every time a label goes on a package. Each of these is a powerful disincentive to cut corners.

The meat industry is the most heavily regulated consumer industry. In addition, proposed changes which would increase industry responsibility are in keeping with the way the Federal government maximizes its resources to protect the public — precisely the way every other product in the supermarket or drug store is regulated. Given the safety record of meat products, and the long history of enjoyment of wholesome meat in this country, we fail to see how anyone can contend that the industry does not, or would not, produce the highest quality products possible.

Reading and Reasoning

EXAMINING COUNTERPOINTS: TOXIC PESTICIDES AND SCIENTIFIC EVIDENCE

This activity may be used as an individualized study guide for students in libraries and resource centers or as a discussion catalyst in small group and classroom discussions.

The Point—by Consumers Union of the U.S.

The mass of existing scientific evidence shows clearly that pesticides are toxic; that several dozen of the hundreds now in use cause cancer in animal tests; that residues persist in some foods and are present in some populations' diets at levels high enough to pose significant theoretical cancer risks, as well as risks of other effects.

Authoritative scientific assessments by the Environmental Protection Agency and the National Research Council agree that pesticide residues pose a substantial public health risk. The EPA, in its 1988 report, "EPA's Unfinished Business," ranked the problems it is charged with solving according to priority. Pesticides in foods was in the group of problems given highest priority—those judged to post "overall high/medium risks." (Congressional Testimony, November 1, 1989)

Counterpoint—by Warren T. Brooks

"The risk of pesticide residues to consumers is effectively zero." This is what some 14 scientific societies, representing over 100,000 microbiologists, toxicologists and food scientists, said at the time of the ridiculous Alar scare. "But we were ignored."

Also ignored by the media was the vast 1967 Total Diet Study by the Food and Drug Administration's Scheuplein, which found that pesticide and fungicide residues were 10,000 times lower than the highest dose shown to cause no toxic effect in the most sensitive lab animals.

As such, the 1967 Food and Drug Administration study should have killed the notion of real risk to humans from agriculture chemicals. (Conservative Chronicle, March, 1990)

Guidelines

1. Which argument do you agree with?

2. Social issues are usually complex, but often problems become oversimplified in political debates and discussions. Usually a polarized version of social conflict does not adequately represent the diversity of views that surround social conflicts.

Examine the counterpoints above. Then write down other possible interpretations of this issue than the two arguments stated in the counterpoints above.

CHAPTER 2

CHEMICALS AND FOOD PRODUCTION

5. THE HEALTH RISKS OF MAN-MADE PESTICIDES   35
 Ralph Nader

6. THE HEALTH RISKS OF NATURE-MADE PESTICIDES   41
 William R. Havender

7. THE DANGERS OF POISON CHEMICALS   47
 Cezar Chavez

8. THE BENEFITS OUTWEIGH THE RISKS   53
 Leonard T. Flynn

9. INTOLERABLE RISK: PESTICIDES IN CHILDREN'S FOOD   59
 Mothers and Others

10. LITTLE PESTICIDES FOUND IN CHILDREN'S FOOD   64
 Food and Drug Administration

11. THE DIOXIN COVER-UP   68
 Peter Von Stackelberg

12. DIOXIN LEVELS ARE NOT HAZARDOUS   74
 Red Cavaney

13. **THE DANGERS OF IMPORTED FOODS: 80
POINTS AND COUNTERPOINTS**
Thomas M. Dorney vs. William L. Schwerner

READING

5   CHEMICALS AND FOOD PRODUCTION

THE HEALTH RISKS OF MAN-MADE PESTICIDES

Ralph Nader

Ralph Nader is a leading national, lobbyist, spokesman and writer for consumer groups and environmental interests. His organization called Public Interest Research Group publishes a national newsletter on consumer issues.

Points to Consider:

1. How is the daily diet of people described?

2. What does the term "acceptable risk" mean for consumers?

3. Define the "clear message" from the Environmental Protection Agency (EPA) and the Food and Drug Administration (FDA).

4. Identify the author's suggestions for consumer groups and industry.

5. What steps should the EPA and Congress take to safeguard the food supply?

Excerpted from congressional testimony by Ralph Nader before the Senate Committee on Environment and Public Works, May 15, 1989.

It is time to realize that pesticides are the silent violence of chemicals and that we need a national policy to phase out their use and embrace non-polluting sustainable agricultural techniques.

In spite of congressional action on a wide range of laws to protect health and the environment over the last 20 years, today virtually every American is exposed in some way to the use of approximately 2.6 billion pounds of pesticides a year. Unlike most other toxic chemicals, pesticides are legally dispersed in the environment and on our food. . .

Daily Diet

While it is not likely that every American's daily diet will contain all of the 300 or more pesticides used on food, it is likely that many will be present on most foods and others have even been found in our water. It is just as frightening that our laws and government allow for such a menu of poison. Many of the older pesticides, still widely used, would probably not be allowed to enter the market if they were developed today. However, they remain on the market through loopholes in the public health protection statutes created for them by Congress.

Today Americans have at least a one in four chance of contracting cancer and a one in five chance of dying from cancer, according to the Department of Health and Human Services (DHHS). The American Cancer Society estimates that one million new cases of cancer will be diagnosed this year. Yet we endure additional cancer risks from pesticides that begin with an infant's first meal. There should be zero tolerance of policies that risk additional cancers given the largely environmentally induced epidemic we are already experiencing. According to a DHHS report, "most scientists now consider that about one third to two thirds of all cancers are associated with the environment in which we live and work."

Acceptable Risk

However, based on the EPA's announced policy of October 19, 1988, the Agency would like to proceed with a policy of "acceptable" risk in regulating pesticides. The implications for continuing a policy that allows cancer causing pesticides in food are major. It means we will continue to see decisions such as the Alar (daminozide) and Captan fiascos which unnecessarily extend the use of carcinogens in the food supply and the environment. A recent study by the U.S. Public Interest Research Group (U.S. PIRG) found that at least 60 pesticides which the EPA believes may be carcinogenic are allowed on commonly eaten foods. The study reported that many of these foods, such as apples, beef, corn, tomatoes, peaches and milk, may legally

"Food stamps don't cover non-food items. I'm going to have to charge you for the chemical additives and pesticide residues."

Illustration by Carol*Simpson

contain 20 or more of these pesticides. Only 11 of these 60 pesticides are in EPA's Special Review process. The Agency seems determined to continue the use of most of these pesticides based on misleading or incomplete information from food and chemical companies.

Until the EPA can prove risk assessment to be more of an exact science than it is today, it should move to ban the sale and use of carcinogens immediately. These chemicals pose more than a risk to the food supply; they endanger people all along the stream of commerce, from production to application. It is unacceptable for the government to defend the use of a cancer-causing agent given the paucity of scientific knowledge about that chemical's use, other chronic health effects, the formula's other ingredients, synergism with other companion pesticides and fertilizers and greater threat to sensitive groups such as infants and children.

All too often, the EPA gerrymanders its decisions about such matters by claiming (with no credibility) to have accurate dietary or pesticide use information that "reduces" the risks posed by a chemical. After nine years in special review, the EPA this year finally acted on Captan. They banned its use on some foods but continued it on high volume foods such as grapes because

> ## 2.6 BILLION POUNDS OF PESTICIDES
>
> *Each year, approximately 2.6 billion pounds of pesticides are used in the United States. Pesticides are applied in countless ways throughout the United States, not just on food crops. They are sprayed on forests, lakes, city parks, lawns, and playing fields, and in hospitals, schools, offices, and homes, and are contained in a huge variety of products from shampoos to shelf paper, mattresses to shower curtains. As a consequence, the pesticides may be found wherever we live and work, in the air we breathe, in the water we drink, and in the food we eat.*
>
> Lawrie Mott and Karen Snyder, Tha Amicus Journal, *Spring, 1988, p. 22*

of its "$1.2 million a year benefits" to the grape and wine industries.

A Clear Message

The message is clear; we can no longer wait for agencies such as the EPA and the Food and Drug Administration (FDA) to protect public health and the environment from pesticides. Their recent statements that "the food supply is safe" after failing to suspend the use of Alar (daminozide) and Captan on apples and other foods were the latest examples of a policy that pits human health and the environment against corporate profit statements.

Although Congress reauthorized the Federal Insecticide, Fungicide and Rodenticide Act (FIFRA) last year, the new law was accurately dubbed "FIFRA Light" in the media because it was so weak. The main improvements made in FIFRA include a 10-year timetable to complete the retesting of approximately 600 active pesticide ingredients and the partial repeal of the indemnification and disposal obligations previously borne by the federal government. With the exception of the testing timetable, the new law will not address many of the problems posed by pesticides today. For example, even the new retesting deadlines fail to mandate comprehensive health testing for neurotoxic effects.

We must therefore act as consumers, as legislators and responsible food and agricultural producers to curb the use of these chemicals today if we are to avoid the "silent spring" that Rachel Carson so ominously predicted 27 years ago. It is time to realize that pesticides are the silent violence of chemicals and that we need a national policy to phase out their use and embrace non-polluting sustainable agricultural techniques. . .

Clearly the resources of the EPA are wholly inadequate given EPA's enormous task of setting food safety limits (tolerance) on more than 8,000 food items for 300 to 400 ingredients in over 40,000 pesticide formulas. Similarly, allocations at the FDA to carry out its responsibility for monitoring and enforcement are as bad or worse. The combined pesticide program budgets ($40 million at EPA and $14 million at FDA) total less than two hours of spending at the Pentagon. Certainly the registration fees added to FIFRA last year need to be enhanced along with new monitoring fees to allow the FDA and the U.S. Department of Agriculture (USDA) to improve both their technological and personnel resources for monitoring and enforcement duties. . .

Suggestions for Consumers

The grape boycott is one of the ways consumers can vote with their shopping carts and send a message to the food industry that pesticides are no longer tolerable on their food. Other useful consumer actions can include supermarket petitions, organizing neighbors to ask store managers to stock organic food and locally grown food in the produce section and encouraging supermarkets to do their own monitoring of products.

Supermarket chains increasingly recognize the demand for pesticide-free food. Consumers should be greatly encouraged to approach store managers about forming an alliance with them. Indeed supermarket customers can form loosely organized committees to meet and discuss new programs with store managers.

Suggestions for Industry

Wholesalers and retailers should begin monitoring of produce and demand that chemical companies provide a practical method to identify pesticide residues. Food retailers should refrain from misleading advertising that promises "pesticide-free food" unless they can certify their claims. These same companies should ask growers to begin phasing out their pesticide use and adopt low-input sustainable agricultural practices. Large food producing companies can also send these messages to growers and adopt these policies on their own farms. Small family farmers may be in a position to profit from a growing consumer awareness and demand for a safer product. Family farmers can convert to organic and sustainable agriculture more easily than monoculture factory farm operations.

Suggestions for the EPA and Congress

- Congress should immediately begin a program to ban carcinogens from the market and seize or order the destruction of existing stocks of banned pesticides, such as Heptachlor, which are still used or allowed in food.

- Congress should oppose any attempt by special interest groups to preempt the right of states to regulate pesticides more stringently than the federal government. All other major environmental statutes allow for the wisdom of the states to lead the way if they choose. If a federal statute is really protective of public health, there will be no reason for states to challenge it.
- The EPA should be required to use the greater risks posed by pesticides to sensitive population groups, such as infants and children, to govern its regulation pesticides.
- The EPA should require pesticide manufacturers to test for more health effects, such as neurotoxic behavioral effects, with special attention given to the potential for learning disabilities and other effects on children.
- The resources of the FDA should be significantly increased to carry out adequate monitoring and enforcement of pesticide residues on domestic and imported food.
- Congress should outlaw the export of banned and unregistered pesticides and insure that users of registered pesticides in receiving countries are adequately notified of the hazards of all pesticides shipped to them from the U.S. application of low-input farming techniques.
- The EPA and FDA currently have inadequately used their authority to notify consumers about pesticides used on their food.

Those engaged in deprecating the seriousness of risks posed by agricultural chemicals and other pesticides to consumers, workers and especially infants may wish to consider the following: If regulation and other internal corporate systems are not strengthened to test, monitor and control the vast array of toxic chemicals now in use, the future will see the emergence of even more reckless efforts to market virulent chemicals in the rabid quest for market share.

READING

6   CHEMICALS AND FOOD PRODUCTION

THE HEALTH RISKS OF NATURE-MADE PESTICIDES

William R. Havender

This report on natural carcinogens in food was written by the late William R. Havender, Ph.D. Dr. Havender was a consultant on environmental carcinogens based in Berkeley, CA, and a Scientific Advisor to the American Council on Science and Health (ACSH). This report was updated by Leonard T. Flynn, Ph.D., a regulatory and scientific consultant.

Points to Consider:

1. How do amounts of nature's pesticides compare with man-made pesticides?

2. How do the hazards of man-made and nature-made carcinogenic substances compare?

3. Are people at serious risk from the danger of natural chemicals in food?

4. What is the best way to minimize potential hazard posed by naturally occurring carcinogens in food?

William R. Havender, "Does Nature Know Best: Natural Carcinogens in America's Food," published by the American Council on Science and Health, April 1989.

Current evidence suggests not only that a normal diet contains substantial amounts of carcinogens and mutagens, but also anti-carcinogens— that is, substances that seem to counteract the action of carcinogens.

Picture this: You glance at a newspaper and notice a headline announcing that laboratory tests have shown that a substance in our food supply is a carcinogen. You only see the headline; you don't have time to read the story. What type of substance would you guess that the carcinogen is?

If you guessed that it was a food additive, a pesticide, or some other type of man-made contaminant, you would have a very good chance of being right. For it is these types of chemicals—man-made substances added to food either accidentally, or deliberately—that make the headlines. However, there is another large group of carcinogenic chemicals that has attracted little attention as yet: naturally occurring carcinogenic substances in food. These natural carcinogens are in fact much more widespread and numerous than man-made carcinogens in food and are present in much larger amounts. In this report, the American Council on Science and Health (ACSH) summarizes the scientific evidence regarding these overlooked carcinogens and discusses their implications for both human health and public safety.

Synopsis

A large number of substances that occur naturally in foods have been found to be carcinogenic (cancer-causing), when evaluated by the criteria customarily used to assess the cancer-causing potential of man-made substances. Many more carcinogens are produced by cooking and by the actions of microorganisms. These natural carcinogens are more numerous, more widespread, and in many cases more potent than man-made carcinogens in food.

It is not necessary or practical for consumers to stop eating foods that contain natural carcinogens. Current evidence does not indicate that these substances have a substantial impact on cancer incidence in the United States. Moreover, natural carcinogens are present in so many foods that it would be unrealistic to attempt to avoid exposure to all of them.

Until recently, scientists and regulatory agencies devoted little attention to natural carcinogens in foods, so much remains to be learned about their potential hazards. However, the information available at this time does not justify concern or efforts to change Americans' eating habits, beyond a recommendation for variety in the diet. Eating a wide variety of foods is desirable for nutritional reasons, and it also helps to reduce the potential hazard posed by naturally occurring carcinogens and other toxins by minimizing the chance that any single substance would be consumed in quantities that would overwhelm the body's defenses. . .

Current evidence suggests not only that a normal diet contains substantial amounts of carcinogens and mutagens, but also anti-carcinogens—that is, substances that seem to counteract the action of carcinogens. These anti-carcinogenic substances include vitamin A and its precursors (particularly beta-carotene), vitamin C, vitamin E, glutathione, purines, and selenium. Laboratory evidence, and in some cases epidemiological evidence, supports the anti-carcinogenic effect of these substances. In addition, butylated hydroxyanisole (BHA) and butylated hydroxytoluene (BHT), two man-made chemicals added to some foods because of their antioxidant properties, have also shown anti-carcinogenic effects in some animal studies.

Health authorities warn, however, that people should not take supplements of these substances in an effort to prevent cancer. Some anti-carcinogens, including vitamin A and selenium, are toxic at levels not much greater than those found in a normal diet. . .

How Do the Amounts of Nature's Pesticides Compare with the Amount of Man-Made Pesticides in Food?

Nature's pesticides are present in foods in much larger amounts than man-made ones. Most of the concentrations mentioned for them are in the high parts per million or even parts per thousand range, while man-made pesticides are present in foods in the low parts per million or parts per billion range. It is estimated that humans typically eat some 10,000 times more of nature's pesticides than of man-made ones. This fact suggests the need to strike a better balance in research efforts in evaluating natural versus man-made pesticides. . .

How does the overall carcinogenic hazard from natural substances compare with that from the man-made substances in food?

We can answer this by looking at two examples of carcinogenic food components of man-made origin. One is saccharin. The other is ethylene dibromide (EDB), the grain fumigant that the Environmental Protection Agency (EPA) banned early in 1984 after residues of it were found in a wide variety of grain-derived foods in supermarkets. EDB is carcinogenic in laboratory animals, and according to the EPA, it is one of the most powerfully carcinogenic of all pesticides.

Saccharin is one of the weakest carcinogens ever detected in

> **"NATURAL CANCER RISK CITED"**
>
> *Natural carcinogens account for more than 98 percent of the cancer risk in the human diet, said Robert Scheuplein, director of the office of toxicological sciences at the Food and Drug Administration (FDA). Even a minor reduction in these hazards would surpass the benefits of eliminating all traces of dangerous man-made chemicals, he said.*
>
> Star Tribune, *February 20, 1990, P.1a*

animal tests. Thus, all of the naturally occurring carcinogens discussed here that have been tested in animals by means of experimental designs that permit comparisons to be made of relative carcinogenic potency with saccharin (that is, administration of the chemical for a lifetime and by the oral route rather than by skin painting or injection) are more powerful carcinogens than saccharin, often far more so.

Aflatoxin B1, for example, is roughly one million times more potent as a carcinogen than saccharin, that is, one million times as much saccharin as aflatoxin B1 would be needed to induce the same incidence of tumors in rats. Thus, a gram of aflatoxin B1 would have the same carcinogenic hazard, based on animal tests, as one million grams of saccharin (about a ton).

EDB, by the same potency scale, is just about in the middle of the range between aflatoxin B1 and saccharin. That is, it is about 1,000 times more potent than saccharin, and about 1/1000 as potent as aflatoxin B1. . .

These examples could readily be multiplied, but they clearly indicate that nature's pesticides, taking into account both their potency and the amounts in which they appear in food, are substantially more hazardous than the man-made pesticides in food. Clearly, research work needs to be balanced more evenly between natural and man-made substances in food.

What about other man-made chemicals in food?

The great bulk of our food—more than 99 percent by weight—consists of the ingredients that Mother Nature put here. Food additives make up less than one percent, and pesticide residues and other man-introduced contaminants (such as from packing materials) cannot even usefully be measured at percent levels: when these contaminants occur, they are present in "trace" amounts. Thus, human exposure to "chemicals" consists overwhelmingly of exposure to chemicals of natural, not man-made, origin. Even a cup of coffee is estimated to contain more than 2,000 natural chemical components, few of which have been adequately studied toxicologically and many of which have never even been identified. At least 150 distinct naturally occurring chemicals have been identified in potatoes, with many more unknown substances also present. Other natural food sources are similarly complex, and the majority of chemical substances contained in them are unidentified. More hitherto unknown substances will be discovered in familiar foods with each advance in analytical techniques.

Moreover, man-made chemicals that appear in the food supply are very tightly monitored and controlled in the U.S. Their use is permitted in foods only at levels that ensure a large margin of safety (typically, at least 100-fold) between the levels of human exposure and the highest level at which no harm of any sort is evident in test animals.

The margins of safety of many natural substances, which we might term "nature's margins," are often much smaller. As mentioned, for example, nature's safety margin for the caffeine in coffee is only about twenty (for a person who drinks five or six cups a day), about ten to twenty for the solanine in potatoes, about ten for the cyanide-generating compounds in lima beans, and about five for salt (before hypertensive effects become evident, although this is also conditioned by genetic factors and disease). Nature's margins for vitamins A and D are about 25 to 40. Indeed, nature's margin for energy intake (calories) is far less than two, for if you consistently eat double the amount of calories you need, you will soon be in trouble owing to obesity and the many health problems related to obesity.

In addition, the toxicological properties of the man-made chemicals in food have been studied far more thoroughly (even while leaving much to be desired) than those of most of the natural components in food. Compounding this lack of knowledge is the fact that comparatively little effort has been put into determining the carcinogenicity of natural compounds, the weight of cancer prevention efforts in the past having been directed toward identifying man-made carcinogens. Thus, when we seriously start screening the natural components of food for carcinogenicity in a systematic way, it is likely that many more carcinogens of natural origin will be recognized. Even now, however, it is apparent (as detailed above) that the known carcinogenic risks from the natural substances in food markedly outweigh those arising from the man-made chemicals. . .

The best way to minimize the potential hazard posed by naturally occurring carcinogens would be to eat a wide variety of foods, since this would minimize the chance that any single carcinogen would be eaten in quantities that would overwhelm the body's natural ability to handle low amounts of hazardous substances with relative safety. Certainly, it would be unrealistic to attempt to remove from our food supply every known trace of naturally occurring cancer causing agents or to avoid all exposure to them. . .

In any case, there is at this point no evidence that low-level exposure to natural or man-made chemicals in the U.S. food supply poses a significant risk of cancer. Obviously, however, if later research should indicate that particularly potent carcinogens are naturally present in certain foods in amounts large enough to justify concern, then actions to reduce our consumption of that ingredient or food would be justified.

READING 7

CHEMICALS AND FOOD PRODUCTION

THE DANGERS OF POISON CHEMICALS

Cezar Chavez

Cezar Chavez is the leader of the United Farm Workers Union and the major spokesman for Spanish American Farm Workers in the American Southwest.

Points to Consider:

1. How many pounds of poisonous chemicals are used in the U.S. everyday?

2. What aspect of farm work has gotten worse in the past 40 years?

3. Identify the five deadly pesticides used in the grape fields.

4. How many pounds of chemical compounds are used in grape production each year?

5. What is the United Farm Workers Boycott?

Cezar Chavez in a national public letter to supporters of the United Farm Workers Union, Winter, 1990.

Each of these deadly poisons has a record of injuring —and in some cases killing—farm workers. And each of these deadly pesticides can be present, as a residue, on the table grapes you buy.

Every day, in the food you eat, you're exposed to chemical residues from over one billion pounds of poisonous pesticides used in the United States every year. . .

Every day, those who work in the fields to produce that food are exposed to over 600 mostly untested chemicals. More than 300,000 farm workers are made ill every year through pesticide poisoning!

A Typical Day

When I was a teenager, just after the Second World War, I had an experience which changed my life. As you read this letter, I hope it will change yours too.

It was a typical day of winter farm work in California's Santa Clara Valley. My brother and I were working as field hands in an apricot orchard, when one day the grower handed us a bucket of strong-smelling liquid, and told us to put a little around each tree.

We started to pour the stuff, and immediately I began to feel nauseated. Since it was right after lunch, I thought it was something I'd eaten, so I walked back to our car. As I walked, I got dizzier and dizzier, and I passed out for several hours.

When I talked to other farm workers, I found out they'd had the same experience. Growers and foremen were telling us it was "medicine" we were pouring and spraying on trees, cotton fields and vineyards. But that "medicine" was giving us skin rashes, making us vomit, causing nosebleeds, and making us weak.

Toxic Pesticides

Many things have changed during the past forty years. But there is one aspect of farm work which has, if anything, gotten worse: the use of powerful, dangerous, toxic pesticides on the fruits and vegetables that you and I eat.

And there is conclusive evidence that you don't have to be a farm worker to be affected. . .seriously affected. . .by chemicals used in agriculture. All you have to do is eat!

This means that you and I are subjected to unknown health risks with every meal. And we're subjecting our children, and generations unborn, to those same risks.

I made a vow, as that teenager sickened by a bucket of tree "medicine", that I would make sure, some day, that neither I nor

Cesar Chavez
P.O. Box 62, La Paz
Keene, California 93531

Source: The United Farm Workers

anyone else would ever again have to risk serious illness through the essential work of providing food.

The United Farm Workers of America, which I helped to organize over 25 years ago, is more than ever committed to achieving that goal: a poison-free environment for all agricultural workers...and for you and your own family.

Campaign

I'm writing this letter because we're in the midst of a campaign which, if successful, will represent a major breakthrough in forcing agribusiness to stop poisoning workers and consumers.

Our tactic is simple. We're asking people to not buy table grapes until the growers agree to three demands:

- The *elimination* of dangerous pesticides from all grape fields...
- A *joint testing program* for poisonous substances in grapes sold in stores...
- *Free and fair elections* for farm workers, and good faith collective bargaining in the grape industry.

We know that, with your help, this grape boycott can work. In fact, it worked before; as you may remember, in 1970 we were successful in eliminating such deadly poisons as DDT, DDE, and Dieldrin from every field under United Farm Workers union contract years before the federal government acted...thanks to a nationwide grape boycott.

Since then, however, the situation has grown even more serious. Thousands of farm workers are poisoned each year in the grape fields. Testing has identified residues of *more than 50 chemical products on grapes sold to you in stores.*

Five Deadly Pesticides

We've identified 5 of the most toxic substances used in the

> ### DEAR MR. CHAVEZ
>
> *Dear Mr. Chavez,*
>
> *I am writing you these few lines to let you know that I am one of those grape workers who now has lung cancer. I believe it comes from when I worked down in the Coachella Valley fields thinning and picking grapes. We were sprayed with some pesticide.*
>
> *We would be working and then told to move over just one or two rows. As soon as the spray tractor passed, we had to go back in the vines and work. We were dripping wet with the chemicals.*
>
> *Last year, after working 17 years, I was told I have lung cancer. I have gotten radiation therapy, hoping and praying I will live a little longer for the sake of my children. I am only 48 years old.*
>
> *May the Lord help you with your work*
>
> *Oralia Rodriguez Valencia*
> *Corona, California*
>
> Reprinted from a United Farm Worker *position paper, 1989*

growing of table grapes. Each of these deadly poisons has a record of injuring—and in some cases killing—farm workers. And each of these deadly pesticides can be present, as a residue, on the table grapes you buy. . .

- **METHYL BROMIDE**. . .extremely poisonous to all forms of life, this fumigant has been responsible for more occupationally related deaths than any other pesticide. Even non-fatal exposure can cause severe, irreversible effects on the nervous system, with permanent brain damage, or blindness. . .
- **PARATHION and PHOSDRIN**. . .can be rapidly fatal, producing illnesses in workers in as little as 20 minutes. Usually sprayed aerially, these poisons cause populations surrounding agricultural areas the same problems as they cause farm workers, since as much as 90% of aerially sprayed pesticides miss their target areas...
- **DINOSEB**. . .poisoning at first resembles heatstroke, then cumulative doses cause extensive illnesses, including loss of vision. It is much too toxic to be used safely (so poisonous, the EPA has finally banned its use, "pending industry reaction."). . .
- **CAPTAN**. . .344,000 pounds are used annually on table grapes, and residue of this compound is the most

frequently discovered material on grapes in stores. Not only can Captan cause cancer, it also causes birth defects and changes in body cells. It is structurally similar to Thalidomide, which caused thousands of babies in Europe to be born without arms and legs. . .

- Not only is each of these deadly pesticides used extensively in the grapes you and I find in stores all across the country. They also have one other thing in common: *each has been recommended for banning by state and federal agencies. . .yet they continue in use!*

It's part of a deadly mass of pesticide poisons used annually in grape production. . .8,000,000 pounds each year of more than 130 chemical compounds.

The most immediate victims of these pesticides are, of course, farm workers. Studies have shown that:

- 78% of Texas farm workers surveyed had chronic skin rashes; 56% had kidney and liver abnormalities; and 54% suffered from chest cavity problems. . .
- The miscarriage rate for female workers is 7 times more than the national average. . .
- More than 300,000 farm workers are made ill every year through pesticide exposure, with grape workers suffering over half of all reported acute pesticide-related illnesses in California. . .

But *you don't have to be a farm worker to be affected* by the pesticides used in agricultural production. . .

Consumers

As a consumer, you should know that. . .

- Pesticides are now thought responsible for groundwater contamination in 23 states. . .and groundwater provides 50% of our country's drinking water supply. . .
- Federal and state pesticide monitoring programs are both flawed and severely inadequate. . .
- In the 1988 harvest season an estimated 380 million pounds of grapes were harvested in Kern County, California, a major grape growing area. Of these 380 million pounds of grapes only 22 samples were taken from the fields. . .
- The samples taken during the actual grape harvest are limited to usually 2 and maybe 3 of the hundreds of toxic pesticides being applied to grapes. . .

Outrageous? Of course it is. And I can't blame you if, like most people, you've felt helpless to do anything about it.

United Farm Workers' Boycott

But now, *there is something you can do!* Something so powerful and effective that if you do it along with us, together we can get those poisons out of our food. . .permanently!

The first step is simple. *Stop buying table grapes!* And if you see them in a friend's house, talk about what you've read in this letter: the poisons used in growing grapes. . .their serious health effects. . .and the United Farm Workers' Boycott.

Then you can further help our effort to get poisons out of the fields. . .and food. . .by asking your supermarket not to promote grapes through advertisements, sales or special displays.

We consider this campaign so important that we have dispatched our entire union leadership to key cities coast to coast. . .not just for a few days, but for as long as it takes to make consumers, stores, and distributors aware that we mean business.

READING

8   CHEMICALS AND FOOD PRODUCTION

THE BENEFITS OUTWEIGH THE RISKS

Leonard T. Flynn

Leonard T. Flynn, Ph.D., M.B.A., is a scientific consultant. He wrote the following report for The American Council on Science and Health. ACSH is a national consumer education association directed and advised by a panel of scientists from a variety of disciplines.

Points to Consider:

1. How is pesticide defined?

2. What are the advantages of using pesticides?

3. Are there any disadvantages?

4. Compare the definitions of insecticides, herbicides and fungicides.

5. Why have bans been placed on useful pesticides?

Leonard T. Flynn, "Pesticides: Helpful or Harmful?", published by the *American Council on Science and Health,* September, 1988.

Pesticides have saved millions of lives in all parts of the world.

A pesticide is any substance that is used by man to control pests. Pesticides are an integral part of modern agricultural production and contribute greatly to the abundance and quality of food, clothing, and forest products our society enjoys. They also protect our health from disease and vermin. Pesticides have often been condemned, however, by proponents of the environmental movement and by some scientists who advocate banning or greatly restricting their use.

Position Statement

- Based on its review of the scientific literature, ACSH concludes that the benefits of pesticides vastly outweigh the risks associated with them.
- Pesticides for home use, agriculture, and health protection have an excellent safety record. When used properly, pesticides do not harm humans or domestic animals.
- Suspension of pesticide use is often made in response to a public outcry and not based on valid or verified scientific data.
- Environmental concerns and potential health risks are often overstated by pesticide critics, while the benefits are improperly disregarded. For example, DDT has saved and continues to protect millions of human lives from malaria and other diseases, yet this very safe and effective insecticide is banned in the U.S. for nonscientific reasons.
- Pesticide critics frequently cite harm to the health of pesticide applicators and farm workers as a prime reason for more controls. The bans of long-used pesticides (e.g., DDT, EDB), however, may have resulted in greater risks to pesticide users due to higher toxicity of the substitutes and the need to use them more often. Replacement pesticides with less well characterized toxicological properties may have to be brought into widespread use when customary and familiar pesticides are forced off the market by ill-advised and excessive regulatory action. Harm to users generally arises from improper and careless use of pesticide products.
- The charge that pesticides "contaminate" America's food is not founded on scientific fact. Much scientific evidence supports the conclusion that traces of pesticide residues in food pose no hazard to human health. . .

Utility of Pesticides

Pesticides have been used for centuries to combat pests. For

Ethyl carbamate is found in naturally fermented foods and beverages, including bread, yogurt, soy sauce, beer and wine. The amounts are small, roughly one to five parts per billion. It produces tumors in a wide variety of tissues in rats and mice, whether administered orally, by inhalation or by injection.

Source: American Council on Science and Health

example, ancient Romans used burning sulfur to control insects and salt to kill weeds. Modern pesticides vary in their uses and are far more efficient than these crude chemical agents. . .

Insecticides control insects that destroy food, clothing and shelter. Crop protection through insecticide use can be spectacular like stopping a plague of locusts, or less visible (but probably more important) like destroying maggots and beetles by fumigating stored fruits and grains. . .

The greatest human health protection resulting from insecticides in recent decades is probably the control of mosquitoes and the many serious diseases they carry, including malaria, yellow fever, encephalitis, dengue and hemorrhagic fevers, and filariasis (elephantiasis). These diseases are not exclusively tropical; both malaria and yellow fever epidemics were reported in North America within the past century. . .

Herbicides are used to control weeds, that is "plants growing out of place." About two-thirds of the volume of agricultural chemicals used in the U.S. are herbicides; insecticides account for about one-fourth. Thanks to the use of herbicides, farmers can greatly reduce competition for water and nutrients by weeds and thereby significantly increase yields of grains, vegetables and fruits.

One authority estimates that the use of herbicides yields at

> **PESTICIDE BENEFITS**
>
> *Pesticides provide three important benefits: increased food and fiber production, improved health protection, and environmental enhancement. They enable us to have an abundant supply of nourishing food at reasonable prices. Pesticides are our first line of defense to avoid food shortage, and they also protect our health by stopping pest-transmitted diseases and sanitizing our food handling and health facilities. Properly used, pesticides improve the environment for plants, domestic animals and human survival, the ACSH report states.*
>
> News Release, American Council on Science and Health, 1988

least five additional bushels of bread grains per acre, which is a 10 to 20 percent increased yield in treated acreage—a net increase of one billion bushels annually, enough for 65 billion loaves of bread, nearly fifteen loaves for each person on earth!

Herbicides also reduce the need for cultivation, saving the farmer fuel and labor while reducing erosion, soil compaction and crop damage which occur whenever farm equipment is driven through fields of growing plants. In fact, herbicides can replace plowing and cultivation entirely through "no-tillage" agriculture. In this way, farmers can greatly reduce erosion and prevent extreme water loss by evaporation from plowed fields. Some sloping land can be used for row crops that otherwise would be subject to wind or water erosion under conventional tillage. Use of no-tillage or reduced tillage agriculture has been increasing, and up to as much as 65 percent of the acreage of crops grown in the U.S. may utilize the no-tillage practice by the year 2000 if present trends continue...

Fungicides help combat various plant diseases. These diseases have been serious problems for mankind's survival throughout history. Wheat rust fungi caused many of history's famines and, until recently, only the development of resistant varieties could combat the rusts. Scientists have now developed fungicides against them, making it possible to control the diseases during the growing season by spraying the crops.

The potato famine in Ireland was caused by the late blight fungus *Phytophthora infestans.* From 1845 to 1851, it destroyed the potato crop, causing an estimated one million people to die of starvation and forcing another million to emigrate.

Prevention of seed decay is critical for good crop yields, particularly when cool weather delays germination. Commercial seed producers use fungicide seed protectants which allow farmers to plant earlier and take advantage of the favorable moisture conditions usually prevailing at the beginning of the season.

Rodenticides control rats and mice in food handling establishments, homes, factories, warehouses and wherever these rodents are nuisances. More than 200 disease-causing microorganisms, parasitic worms and insects are associated with rats and mice, including plague, leptospirosis and murine typhus. Rats plunder one-fifth of the world's crops each year and are accurately called "man's worst enemy". . .

Nematicides control nematodes, small hair-like worms, many of which live in the soil and feed on plant roots. Practically all pineapples must be treated with a nematicidal fumigant, and soybeans, even those resistant to nematodes, can produce much higher yields with the help from nematicides in nematode infested soil.

Antimicrobial sanitizers and disinfectants reduce the number of pathogenic organisms to non-hazardous levels on hard surfaces like floors, walls and countertops in food service establishments and hospitals. Pharmaceutical, medical device, food and cosmetic manufacturers also use hard surface disinfectants for general sanitation to control product contamination by microbes. . .

Scientific Knowledge

With the development of increased scientific knowledge plus the use of modern pesticides and fertilizers, the past 40 years have brought more progress in agricultural production than in all previous recorded history. Pesticides have saved millions of lives in all parts of the world due to disease vector control and hygiene programs. Nevertheless, public "chemophobia" — the unreasonable fear of chemicals — has led to bans on useful products and has jeopardized this progress. Despite the enormous improvements in living standards, it seems "mankind still finds new things to make himself miserable." The news media and other groups too often "sensationalize dangers" and fail to provide a "meaningful perspective" on pesticides and pest control issues. As a result, opinions can become polarized. One report by the National Research Council summarized the situation as follows: "Users of pesticides fear that they will be regulated to the point where pests cannot be effectively controlled, with concomitant losses of food while opponents of the use of pesticides fear that people are being poisoned and that irreversible damage is being done to the environment."

Pesticide Fears

The environmental and health fears of pesticide opponents appear groundless. In contrast, the concerns of pesticide users that their livelihoods may be jeopardized by bans or other severe limitations do not seem to be unjustified based on recent events.

Even nonchemical methods of pest control can be frustrated by nonscientific and political attacks. Genetic modifications ("gene-splicing") to create new microbial insecticides and improve crop plants have been tied up by legal maneuvers despite considered scientific opinion that "there is no evidence that unique hazards exist" for such substances. Similarly, the use of irradiation to replace chemical fumigants has been slowed by regulatory obstacles.

Pest Control

Ideally, the main thrust of regulation, science and politics should be to improve the methods of pest management. None of our pest control systems is perfect, and because the pests keep evolving, our present techniques may be even less effective in the future. Research and development on a wide variety of fronts must continue in order to stay even and in hopes of pulling ahead. This means encouraging research to develop better pest control tools, including safer and more effective pesticides. Prudence on all sides—environmentalists, industry, researchers, and regulators—would be welcome.

Science and scientists must not be brushed aside by hysteria and the destructive political decrees which follow. We must not forget that despite the fears and real problems they create, pesticides clearly are responsible for part of the physical well-being enjoyed by most people in the United States and the western world.

READING

9   CHEMICALS AND FOOD PRODUCTION

INTOLERABLE RISK: PESTICIDES IN CHILDREN'S FOOD

Mothers and Others

Mothers and Others for Pesticide Limits, a project of the Natural Resources Defense Council, is working nationally to call attention to the problem of pesticides in children's food. The Natural Resources Defense Council (NRDC) is a nonprofit membership organization dedicated to protecting natural resources and improving the quality of the human environment. NRDC has more than 100,000 members and is supported by tax-deductible contributions.

Points to Consider:

1. Explain the message of the two-year study by the NRDC titled, "Intolerable Risk: Pesticides in Our Children's Food".

2. What is the nature and names of four pesticides that pose the greatest risk to children?

3. Identify the loopholes in the way pesticides are currently being monitored by the EPA and regulated by the FDA.

4. What are suggested solutions to this problem?

5. How can parents minimize pesticide food risks to their children?

Excerpted from a public paper by the National Resources Defense Council titled, "Our Children Are in Danger", 1989.

The average child receives four times more exposure than an adult to eight widely-used carcinogenic, or cancer-causing, pesticides in food.

The fruits and vegetables that we feed our children contains nutrients that are crucial for their growth. Unfortunately, that's not all they contain. Today, because of the widespread use of pesticides, fruits and vegetables also contain dangerous pesticide residues that may be threatening our children's health.

The Natural Resources Defense Council (NRDC) recently completed the most comprehensive study ever undertaken on this problem. The groundbreaking two year study, "Intolerable Risk: Pesticides in Our Children's Food," has uncovered startling information about the dangers of pesticides and how they are regulated.

Most alarmingly, the study found that **children are exposed to more pesticides in their food than adults are—at a time in their lives when they may be especially vulnerable to pesticides' harmful effects.** In fact, NRDC found that:

- The average child receives four times more exposure than an adult to eight widely-used carcinogenic, or cancer-causing, pesticides in food.
- As a result of this exposure to only eight pesticides during their preschool years alone, as many as 6,200 children may develop cancer at some time in their lives. The Environmental Protection Agency has identified 66 of the 300 pesticides used on food as known or suspected carcinogens.
- At least 17 percent of the current preschool population—or 3 million children—may be exposed to "neurotoxic pesticides, which can cause nervous system damage, at levels above what the government considers safe.

There's a simple reason for these problems: **The government is not doing its job of protecting children against pesticides in food.**

"Legal" Food Doesn't Mean Safe Food: The Regulatory Failure

According to the NRDC study, there are a number of dangerous loopholes in the way pesticides are currently being regulated by the Environmental Protection Agency (EPA), and the way pesticide limits are being monitored and enforced by the Food and Drug Administration (FDA). **In effect, in regulating pesticides in food the government virtually ignores their impact on children.** Specifically, the NRDC study found that:

"After the latest CANCER scare, we thought you needed a little cheering up."

Illustration by Carol*Simpson

- **The EPA did not take children's eating patterns into account** when it set virtually all current legal limits on pesticide residues in food. Instead of considering what and how much food children eat, the EPA has relied on estimates of adults' food consumption for the vast majority of residue limits now in effect. Since children actually eat proportionally more food, per body weight, than adults do, they're also consuming proportionally more pesticide residues. And children's diet's are dominated by fruit, which is more likely than other food to be contaminated with pesticides.
- **The EPA doesn't adequately take into account children's susceptibility to hazardous pesticides.** Children may be especially vulnerable to carcinogens, which can cause cancer, and to neurotoxins, which can harm the nervous system. But the EPA doesn't require enough testing of pesticides to determine their particular effects on children's neurological and behavioral development.
- **FDA sampling and testing for illegal pesticide**

SOME OF THE WORST PESTICIDES

Here are some of the pesticides that the NRDC found to pose the greatest risks to children:

Daminozide *is sprayed on apples and other fruits to improve their appearance, increase their shelf life, and make harvesting easier. It breaks down the UDMH when the fruit is processed into products like applesauce and apple juice. UDMH is a potent carcinogen.*

Mancozeb *is a fungicide used primarily on tomatoes, potatoes, and apples. During processing it breaks down the ETU, which is carcinogenic and otherwise toxic.*

Captan *is a carcinogenic fungicide used on strawberries and other fruits.*

Methamidophos, parathion, methyl parathion, and diazinon *are all "organophosphate" insecticides that can cause nervous system damage.*

contamination is inadequate. Even the general public may not be protected against unsafe pesticide residues, since the FDA samples only about one percent of the food supply, and since its routine testing can only detect about one half of the pesticides that could be present in food. What's more, there are so many delays in FDA laboratory testing of food that by the time illegal pesticide contamination of food is detected, the food has probably been sold and eaten.

There Are Solutions

There are several things that you can do—as a parent, as a consumer, and as a citizen—to start protecting your children from pesticides in food. These include some short-term steps you can take right away, like buying and preparing produce carefully to reduce the likelihood of pesticide exposures. **But what's needed in the long run will be important reforms in the way our government regulates pesticides, as well as some basic changes in the way our country's food supply is grown.** Here are some of the most important changes that are needed:

- Congress must require the EPA to protect children from pesticides in food by revising its legal limits on pesticide exposures to take into account children's higher consumption of produce.
- The EPA should also be required to take pesticides off the market if they're found to pose serious risks.

- The Food and Drug Administration must improve its methods for sampling food and detecting pesticide residues.
- The federal government should provide farmers with financial assistance and other incentives to switch to safer, low-pesticide farming techniques. These techniques, such as organic farming and "integrated pest management", have been used successfully for years, and could significantly reduce the dangerous pesticides in the country's food supply. Because of the high costs of increased pesticide use—both economic and in terms of health—many farmers would probably switch to safer techniques if it weren't for the initial financial risks involved.

What's a Parent to Do?

1. **Write to government and elected officials.** Write to the Environmental Protection Agency, the Food and Drug Administration, and your U.S. Senators and Representatives, to press for the above reforms.

2. **Meet with your supermarket manager.** Urge him or her to start stocking organically-grown produce, and to label imported produce (which often contains more pesticide residues than domestically-grown produce).

3. **Wash produce carefully, and peel it when appropriate.** Washing produce in a diluted solution of water and dish detergent may remove some of the surface pesticide residues. (Be sure to rinse well!) Peeling produce will remove all surface residues. Unfortunately, though, peeling also removes some nutrients—and neither peeling nor washing will remove pesticides that are contained inside the produce.

4. **Grow or Buy organically-grown produce.** Certified organically-grown produce usually means that it was grown without the use of synthetic fertilizers, pesticides, or growth regulators. If you buy organically-grown food, you're helping to protect your family, and also exercising your power as a consumer to show grocers and farmers that food contaminated with pesticides is unacceptable.

5. **Join NRDC's Mothers and Others for Pesticide Limits campaign.** You can join us, and you can also order *For Our Kid's Sake*, which is packed with practical solutions for the problem of pesticides in children's food.

READING

10 CHEMICALS AND FOOD PRODUCTION

LITTLE PESTICIDES FOUND IN CHILDREN'S FOOD

Food and Drug Administration

The following statement was excerpted from a public paper by the Food and Drug Administration on the issue of pesticides in children's food. The FDA is responsible for keeping the food supply safe through its testing and watch-dog efforts.

Points to Consider:

1. Who is responsible for registering and approving pesticides for use?

2. Compare the functions of the EPA, the FDA and the USDA.

3. What has the FDA found in its testing and sampling of baby food in the past 25 years?

4. How is the FDA's "Total Diet Study" explained?

5. What is Alar and why has its use been halted?

6. How is the term "tolerance" defined?

Food and Drug Administration, "Talk Paper" on Children's Food, February 27, 1989.

FDA for many years has conducted surveys specifically of baby foods—infant and toddler or junior foods. Of the several thousand samples analyzed during the past 25 years, only about one-fourth had detectable residues—none of which exceeded EPA's legal limits.

The Natural Resources Defense Council Inc.—an advocacy group on environmental issues—has issued a report alleging that the Environmental Protection Agency does not take the exposure of children and infants sufficiently into account when developing pesticide tolerance. While questions about tolerances and their safety margins should be referred to EPA (which is responsible for registering and approving pesticides), FDA for its part can assure the public that the residues actually found in foods—and in infant and toddler foods in particular—are generally far below these EPA tolerances.

Twenty Five Years

FDA has been monitoring foods for pesticides and other contaminants for more than 25 years. The agency is responsible for enforcing EPA tolerance levels for foods in interstate commerce, except meat and poultry which are a U.S. Department of Agriculture responsibility. FDA monitors residues (along with USDA, the states, food manufacturers and private organizations) to see that they remain within EPA tolerances. FDA's test results were detailed in a report late last year. (See Press Release P88-37, Dec. 2, 1988—"Contrary to common belief, most FDA tests show little or no detectable pesticide residues.")

Special Surveys of Baby Foods

In addition to the general sampling and testing of raw and processed domestic and imported foods reviewed in the report and press release, FDA for many years has conducted surveys specifically of baby foods—infant and toddler or junior foods. Of the several thousand samples analyzed during the past 25 years, only about one-fourth had detectable residues—none of which exceeded EPA's legal limits. The samples included infant formula, cereal, fruit and fruit juices, vegetables and desserts for children.

Total Diet Study for Children, Teenagers

FDA also conducts a Total Diet Study each year: FDA buys 234 food items from supermarkets in each of four regions, prepares them (washing and cooking as appropriate) and analyzes them for residues of pesticides and other contaminants.

Reprinted by permission: *Tribune Media Services.*

Actual daily intakes of the residues found are calculated for various age groups, including children 6 to 11 months, and 2 years, and for teenagers 14 to 16. These show that:

- Pesticide residues are consistently below EPA's tolerances and the dietary intakes thereof and are but a fraction of the UN's Food and Agriculture Organization's and the World Health Organization's acceptable daily intakes. The diets of children have consistently shown less than one percent of the applicable acceptable intakes of FAO/WHO for most pesticides.

- Of nearly 800 samples of commercial baby foods analyzed as part of the Total Diet Study since 1982, the majority showed no residues at all. Of the 43 percent that contained measurable residues, the highest average level was for permethrin at an amount about 1/30th the EPA tolerance.

Declining Use of Alar

A chemical that has been of concern to the public, especially in the case of children, is daminozide (trade name Alar). On the basis of results from still ongoing animal studies of potential carcinogenicity, EPA recently announced that it plans to halt use of the product. Once used on 40 percent of apples to increase

> **RISK ESTIMATES**
>
> *Risk estimates for Alar and other pesticides based on animal testing are rough and are not precise predictions of human disease. Because of conservative assumptions used by EPA, actual risks may be lower or even zero, the government said.*
>
> John Nicholson, Human Events, May 13, 1989, p. 10

their firmness, enhance color and extend storage life, Alar currently is applied to perhaps 5 percent of the apple crop. After questions were raised about daminozide's use on apples, major baby food manufacturers two years ago required their suppliers to abstain from its use.

Analyses of fruit samples, including applesauce and apple juice for children, show either no daminozide residues or levels well within EPA tolerances. For example, when daminozide's use was more widespread, FDA in 1986 analyzed apples, applesauce, apple juice and other products for daminozide and its breakdown product, UDMH. None of the apple product samples exceeded 1 ppm daminozide or 0.07 ppm UDMH. The EPA tolerance for daminozide is 20 ppm for apples and apple products. The highest level found was 5.9 ppm daminozide in canned cherries, another permitted use. Tolerance for this food is 55 ppm. FDA's monitoring programs are continually modified to expand coverage, reflect new pesticide use patterns or answer newly raised questions. For example, FDA is negotiating an agreement whereby milk collected by EPA nationwide (for other purposes) would be supplied to FDA for pesticide analysis, thus giving additional coverage for this important food.

Tolerance

A tolerance should not be taken to mean the maximum safe level of a pesticide, and that anything over the tolerance is not safe. The tolerance may be considerably less than the maximum safe level, EPA reports, because EPA sets a tolerance no higher than the highest residue that should result from a pesticide's intended use. Nor do tolerances themselves represent the levels normally found on foods but the maximum legal residue; actual residues are much lower. (See "Setting Safe Limits on Pesticide Residues" in the October 1988 FDA Consumer magazine.)

The acceptable daily intake is the amount of a pesticide which, if consumed over a lifetime, appears to be without appreciable risk to the health of the consumer.

READING

11 CHEMICALS AND FOOD PRODUCTION

THE DIOXIN COVER-UP

Peter Von Stackelberg

Peter Von Stackelberg is a freelance writer living in Edmonton, Alberta. He wrote the following article for Greenpeace *magazine.*

Points to Consider:

1. Describe the dangers of dioxin.

2. How is dioxin defined?

3. What has the paper industry done about the dioxin problem?

4. Compare the reactions of Americans and Europeans to the dioxin problem.

5. How has the Environmental Protection Agency (EPA) reacted to the dioxin problem?

Peter Von Stackelberg, "White Wash the Dioxin", *Greenpeace,* March-April, 1989, pp. 7-11.

TCDD is also the deadliest substance ever produced. Its toxicity has been compared to plutonium.

Since at least 1980, the Environmental Protection Agency (EPA) scientists and researchers with Canada's environment and health departments have been expressing their concern about the growing dioxin contamination of the environment. They are concerned about the high toxicity of dioxin and its extreme ability to bioaccumulate. Dioxin is the term commonly used to describe a group of about 75 compounds with the same basic chemical structure.

2,3,7,8-tetrachlorodibenzo-p-dioxin (TCDD) is the most studied member of the dioxin family. TCDD is also the deadliest substance ever produced. Its toxicity has been compared to plutonium—the EPA's procedures for handling these two materials are the same.

Dioxin Dangers

Industry representatives have argued that low levels of dioxin do no harm, but this contention has never been supported by scientific research. During congressional hearings in 1980, EPA scientists testified that TCDD was so powerful a carcinogen and teratogen that even the lowest measurable doses caused cancer and birth defects during laboratory tests. "EPA considers dioxin a carcinogen and, as all carcinogens, considers there to be a finite risk at any level," said the EPA's Alec McBride. "EPA considers any level as posing a degree of risk."

Cancer is not the only danger that dioxin poses. The effects of TCDD in all species of animals tested under laboratory conditions included weight loss, liver damage, hair loss, abnormal retention of body fluids and suppression of the immune system. Other effects of exposure to TCDD include birth defects and infertility. The dangers to the unborn in particular were emphasized by the EPA's Don Barnes. In a memo written on March 16, 1987, he said:

Mothers and Infants

"Pregnant women, lactating mothers, developing fetuses and nursing infants constitute a subpopulation of special concern. Human body burdens [of dioxins and furans, a closely related group of highly toxic chemicals that are often found with dioxins] are likely to lead to additional burdens to the fetus and the nursing infant, which are not mimicked in the animal tests. Increases in the mother's body burden as the result of [dioxin/furan] contaminated food would likely lead to additional exposures."

In the late 1970s and early 1980s, public concern over dioxin contamination centered on sites like Love Canal and Times

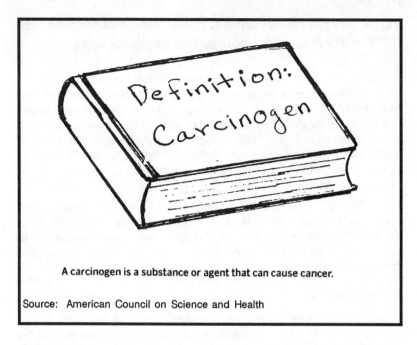

A carcinogen is a substance or agent that can cause cancer.

Source: American Council on Science and Health

Beach. But it soon became evident that dioxins are far more widespread in the environment than two places in New York and Missouri...

Public Strategy

In addition to forestalling regulatory action, the paper industry put together a "dioxin response team," which recommended "a public affairs strategy calling for activities keyed to, and in advance of, the release of the joint EPA/industry dioxin study."

A comprehensive API plan dated March 2, 1987, treated the public health threat posed by dioxin as a public relations problem. The industry's strategy was to: "1) keep all allegations of health risks out of the public arena—or minimize them; 2) avoid confrontations with government agencies, which might trigger concerns about health risks or raise visibility of issues generally; 3) maintain customer confidence in integrity of produce; and 4) achieve an appropriate regulatory climate."..

Although the pulp and paper industry was trying to stifle the release of government information, it discovered some disturbing new revelations about the dangers that dioxin contamination of paper products presented. TCDD, the most deadly of the dioxins, was found in bleached pulp at levels ranging from one part per trillion to 51 parts per trillion (ppt) in the vast majority of the samples taken. Levels of related chemicals were found to range from 1.2 ppt to 330 ppt.

> **CLEAN, WHITE AND DEADLY**
>
> *In October, Greenpeace campaigner Renate Kroesa appeared on Canadian television surrounded by an array of tan paper products—coffee filters, milk cartons, diapers and toilet paper—purchased on a shopping spree in Sweden. What makes these items different, and accounts for their unusual color, is the fact that they are made from unbleached paper...*
>
> *Kroesa's TV appearance comes on the heels of a study by the Health Protection Branch in Ottawa showing that dioxin from bleached paper milk cartons has migrated into Canada's milk.*
>
> "Clean, White and Deadly", Greenpeace, *January-February, 1989, p. 4*

Public Danger

Part of the industry's public relations strategy was to dismiss these levels—"trace amounts" as they have often been called—as being far below any level that presented danger to the public...

Yet other tests sponsored by the API itself showed that superabsorbant disposal diapers had up to 11 parts per trillion of dioxin in them, paper towels up to 7 parts per trillion, and various types of paper plates up to 20 parts per trillion. A draft report prepared for the industry in June 1987 by the research firm of A. D. Little found that between 50 and 90 percent of the dioxins in paper products "in contact with food, oils, or water is available for consumption."..

Although EPA is aware of the connection between pulp mills and dioxins, the agency had failed to produce regulations that would eliminate dioxin contamination of air, water and paper produces. Instead, in April 1988, the EPA decided to do another, bigger study of all 104 pulp and paper mills that use chlorine...

The Environmental Defense Fund and National Wildlife Federation sued EPA for its complacency. Settling out of court, EPA agreed to complete a risk assessment of the 104 mills by April 30, 1990. But even after that is done, EPA can: 1) refer the problem to another federal agency; 2) decide that dioxin from bleached wood pulp doesn't produce an unacceptable risk; or 3) take another year (until April 30, 1991) to propose regulations.

Europe

While North America studies dioxin, several European governments have decided to deal with the problem head on. Throughout Europe, the need for highly bleached paper products is being re-evaluated. Sweden, for example, has stopped the sale of chlorine-bleached disposable diapers. In Austria, consumers are using unbleached brown coffee filters and milk cartons.

"Household products are one of the most important keys in the struggle against environmentally unsound consumer goods," said Brigitta Dahl, Swedish minister of the environment. "Therefore we are now concentrating our efforts against chlorine and dioxins in the most common household products. This will be a strategically important contribution. If one gets paper bleached with chlorine out of consumer products, one also gets large amounts of chlorine out of the industrial stage, and the consumers don't have to live with an environmental threat on their breakfast tables, in their bathrooms and in large parts of their lives."

Sweden is making great strides in getting dioxins out of its pulp and paper mills. Today the Swedish pulp industry discharges about 3.5 kilograms of organically bound chlorine per ton of pulp. But new laws require that mills reduce their discharge to 1.5 kgs/ton by 1992 and completely stop it by the year 2000. Swedish mills are using oxygen bleaching, among other things, to meet this goal.

North America

The North American pulp and paper industry has used delaying tactics to avoid legal liability for medical problems that people may have suffered as a result of exposure to dioxin, Van Strum said. She says the EPA is hesitant to regulate dioxins for the same reason.

"There is no question they are trying to avoid regulating the most toxic known substance. There are two reasons for this," she said. "First, if the EPA were to say dioxins were hazardous, they [would] create a liability in cases like the Agent Orange litigation. Second, several laws the EPA administers specifically state only safe doses of pollutants can be released into the environment. Saying dioxins are hazardous at any level would seriously affect many industries and activities."

Inadequate Solutions

Industry recognizes dioxin-contaminated paper as a problem. Unfortunately, some solutions being offered are inadequate, because they eliminate or reduce dioxins in the final product, but continue releasing them into the environment. For example, Dow Chemical Corporation is developing ion-exchange resins that would remove organochlorines from pulp. But if this method were implemented, the dioxin-saturated resins would still have to be disposed of somewhere.

A major political fight may be required to stop the spread of deadly dioxins and other organochlorines. Regulations that include a timetable for zero discharge of these toxics seem far away. EPA policy makers continue to stall; they attempted to reduce their obligation to clean up dioxin-contaminated sites by announcing last year their intent to increase—by 16 times—the levels of exposure that will be deemed acceptable. Fortunately, EPA's Science Advisory Board agreed at their December 1988 meeting that "there is no firm scientific evidence" for the proposal.

Greenpeace is pushing for standards that will completely eliminate organochlorine discharges by 1993. This can be achieved by abandoning the use of chlorine in the bleaching process. "What is needed is a lot of local participation by people," Paul Merrell says. "That is the only way that the spread of dioxins into the environment from pulp mills will be halted quickly," he adds.

Changing Demand

In addition to political pressure, the economic weight brought by changing consumer demand for bleached paper products may be needed to force government and industry in North America to deal with the problem. Greenpeace has asked for the immediate introduction of chlorine-free and/or unbleached paper products, as well as a higher recycling rate of paper products. The North American industry has resisted these demands. Coffee filter producers, for example, claim that they do not have enough unbleached pulp to produce unbleached filters. Yet at the same time the pulp industry is undergoing an enormous expansion program, all geared to producing more chlorine-bleached pulp. Unbleached pulp is cheaper and easier to manufacture.

Hygiene

For years whiter-than-white paper products have been associated with hygiene by consumers. Now they should be seen as a threat to health and the environment. "Paper is a natural product, made of a potentially renewable resource," says Renate Kroesa. "How can we ever come to terms with living on this planet if we don't even produce paper in an environmentally sound way?"

READING

12 CHEMICALS AND FOOD PRODUCTION

DIOXIN LEVELS ARE NOT HAZARDOUS

Red Cavaney

Red Cavaney is President of the American Paper Institute. The 175 companies who hold membership in API manufacture over 90 percent of the pulp, paper and paperboard produced in the United States. The industry operates over 500 paper mills in 42 states, directly employing 246,000 people.

Points to Consider:

1. Why are paper products safe from any dioxin threat?

2. How did the problem of dioxin originate?

3. What levels of dioxin did the Environmental Protection Agency (EPA) find in consumer products?

4. How has the paper industry reacted to the dioxin issue?

5. What trace levels of dioxin did American Paper Industry scientists find in consumer products?

Excerpted from testimony by Red Cavaney before the House Committee on Energy and Commerce, December 8, 1988.

Extensive scientific analysis has confirmed that bleached paper products are safe, as is the workplace in which they are manufactured.

I am pleased to appear before you today to discuss how the industry has responded to the unexpected discovery of previously undetectable trace amounts—in parts per trillion and quadrillion—of dioxin in association with some of the industry's bleached pulp and paper mills and in some bleached paper products. Since that initial detection, understanding dioxin—how it is formed, its potential effect on health and the environment, and how to substantially reduce its formation—has been a top priority of the paper industry. It is an effort to which the members of API and the industry's separate research organization NCASI* have committed significant resources. This has included using the best science available—often pushing the boundaries of existing technology—to get to the bottom of this matter and pursuing the issue on an expedited basis. We are equally committed that the effort be conducted in the open, with test results and conclusions provided to interested parties as they are confirmed.

Paper Products Are Safe

At the outset, I wish to assure the Committee that extensive scientific analysis has confirmed that bleached paper products are safe, as is the workplace in which they are manufactured. In the testing completed to date of the industry's workplace, its products, effluent and sludge, dioxin (TCDD) and furans (TCDF)—a related compound with one-tenth the potency of dioxin—have been detected only at trace levels ranging from parts per quadrillion to low parts per trillion. In fact, science's ability to detect the presence of any dioxin or furans at these levels has only recently become possible because of extraordinary advances in detection instrumentation and analytical capability.

More importantly, comprehensive assessments (later to be described herein) clearly demonstrate that even when these minute amounts are found, the products and workplace are safe.

I would like to describe how this became an issue for the pulp and paper industry.

---

*NCASI, the National Council of the Paper Industry for Air and Stream Improvement, is the industry's environmental technical and research organization.

Reprinted with permission: *Tribune Media Services*

Origin of Problem

As you have heard from previous witnesses, approximately 38 months ago the U.S. Environmental Protection Agency found traces of dioxin in individual samples of pulp and paper industry sludge during the course of its Congressionally-mandated National Dioxin Study. Although the dioxin found by the Agency could only be measured in the low parts per trillion range, we in the industry, as well as EPA and state agencies, were surprised at this discovery. Adding to this surprise was the fact that in 1984 the industry had, in cooperation with the State of Maine, tested mill sludges and found no traces at all of dioxin, utilizing what were then state-of-the-art testing and analytical techniques.

The trace amounts of dioxin found by EPA were detected with scientific instrumentation and analytical techniques that are 1,000 to 1,000,000 times more sensitive than the methodology previously available. The new emerging methodology became available in 1985, and even today only a small number of labs in the United States are capable of utilizing it. The development of more sophisticated detection capabilities explains why dioxin had never before been associated with the papermaking process—a basic process that has remained fundamentally unchanged for decades.

Actions by Industry

When the industry learned of the trace amounts of dioxin, we took immediate action. Within a week of the original detections,

> ## LOW DIOXIN LEVELS
>
> *The Food and Drug Administration has confirmed earlier reports that extremely low levels of dioxin migrate from bleached paper cartons into the milk they contain, but in a letter informing Congress of its findings, FDA said today that the levels measured well below one part per trillion...*
>
> *FDA also said paper manufacturing techniques are being changed, and that the very low levels do not pose a threat during the time it takes to complete the changes.*
>
> *"During the short period of time it will take to complete corrective steps, milk is safe to drink," FDA Commissioner Frank E. Young, M.D., Ph.D., said. "But because we have the means to virtually eliminate even this low level of dioxin, it is prudent to do so."*
>
> News Release, U.S. Department of Health and Human Services, September 1, 1989

and non-stop ever since, the industry has been engaged in a vigorous and extensive examination to answer the following five questions:

- How is the dioxin formed and in what amounts?
- Is it in consumer products resulting from the pulp bleaching process and, if so, does it pose any health risk to consumers?
- Is there any impact in the workplace or on our employees?
- What are the effects on receiving streams and their consumable fish population?
- And, finally, can the minute amounts of dioxin found be reduced?

In an attempt to answer the first of these questions—how is the dioxin formed and in what amounts—we undertook a joint screening study with EPA to examine five geographically dispersed paper mills which volunteered to be tested. The goal of the study was to develop, with the help of the best qualified lab, data in which both industry and EPA could have confidence.

Public Release

The results of the five-mill study were released publicly by EPA and the industry last year. This is what we found.

In effluent from three of the five mills, we found dioxin in low parts per quadrillion. Two mills had no detectable dioxin.

In bleached pulp from four of the five mills, we found dioxin in the low parts per trillion. One mill had no detectable dioxin in pulp.

Finally, in sludge we found dioxin at all five mills in the low parts per trillion.

Just for perspective, 1 part per trillion can be viewed analogously as equal to one second in 32,000 years, and 1 part per quadrillion is equal to one second in 32 million years. Put another way, 1 part per trillion is equivalent to less than the thickness of a credit card compared to the distance to the moon. . .

Major Investigation

These findings led us to undertake a major investigation as to how dioxin was being formed. But even as preliminary results began to confirm the presence of trace levels of dioxin in some bleached pulps, the industry moved very quickly to address the second question: Is it in consumer products and does it pose any health risk? . .

The results to date have been extremely encouraging and have been reviewed by the three agencies of the federal government represented here today. Due to the low migration rate and the patterns of consumer use, it is critical to note that the levels of dioxin and furans found to date in products tested are lower—in many cases tens of thousands to millions of times lower—than the levels at which the FDA, as an example, would have any concern. . .

The third question we posed related to the workplace. Here we wanted to know if inhalation of paper dust containing any traces of dioxin, or worker contact with bleached paper, would pose any problems. The risk assessments completed and made public earlier this year clearly indicate there is no apparent dioxin-related risk posed to workers through paper dust inhalation. Similarly, worker handling of bleached paper does not pose a health concern, with a wide margin of safety.

The fourth question related to streams and fish. What is the impact of the small amounts of dioxin remaining in the treated effluent from some of the industry's bleaching mills on receiving streams and the nearby consumable fish?

The industry—both through NCASI and individual companies' efforts—has for many years conducted extensive studies on the impact of treated pulp mill effluents on receiving streams. These studies have shown no significant adverse impact on the aquatic health of those streams. . .

Even though the testing to date has confirmed the safety of our products and the workplace, we are continuing to address the final question: Are there ways to reduce dioxin formation when it does occur in our manufacturing process? . .

Currently, none of us knows of a commercially available technology which would totally eliminate dioxin formation. I underscore this because continuing advances in detection technology may make "total elimination" a false promise...

The industry is committed to utilizing a variety of commercially available manufacturing process options which will lead to a fundamental overall reduction in dioxin formation associated with bleached pulp and papermaking...

Sweden

It may be appropriate at this point to respond to recent questions that have been prompted by some confusion over comparisons between the U.S. paper industry's efforts and those of the Swedish pulp industry. There is a major difference between these two industries. Effluent from roughly two-thirds of the pulp produced in Sweden does not receive secondary, or biological, treatment before being discharged into receiving waters—primarily the Baltic Sea. Governmental, industry and citizen concern over pollution in the Baltic, particularly with respect to chlorinated organics, led to calls for less use of chlorine as a pulp bleaching agent. In this vein, some consumer products are now being offered and promoted in Sweden as not being bleached with chlorine. This was, and continues to be, a response to a shared *environmental* concern over effluent—not to a *product safety* concern. In contrast, virtually all U.S. craft mills biologically treat their effluent, which removes significant portions of the chlorinated organics...

Much Work Remains

We still have much work ahead of us. This remains a total industry effort. We are optimistic about meeting the challenges we still face, and we intend to apply the same intensity of effort to these remaining questions as was directed at this issue from the very beginning.

As such, we remain vigilant in our commitment to quality products and a safe employee and community environment. I will repeat that as we go through this process, API is equally committed to keeping interested parties informed of the results of the industry-sponsored research efforts.

READING

13 CHEMICALS AND FOOD PRODUCTION

THE DANGERS OF IMPORTED FOODS: POINTS AND COUNTERPOINTS

Thomas M. Dorney vs. William L. Schwerner

Thomas M. Dorney is a special research assistant to the House Committee on Energy and Commerce. William L. Schwerner is the Assistant Associate Commissioner for Regulatory Affairs in the Food and Drug Administration. In the following counterpoint exchange, the dangers of imported food are debated.

Points to Consider:

1. How would you summarize the position taken by Thomas Dorney?

2. What does he say about the use of form 701 in food import enforcement?

3. Where does he say rejected food imports end up?

4. Explain how William L. Schwerner's position on the safety of food imports differs from that of Mr. Thomas Dorney's.

5. How does the Food and Drug Administration (FDA) examine imported foods?

Excerpted from testimony by Thomas M. Dorney and William L. Schwerner before the House Committee on Energy and Commerce, September 28, 1989.

Thomas M. Dorney

In December 1988, the United States Customs Service (Customs) proposed that its inspectors stop collecting Food and Drug Administration Form 701, the key document in the FDA's import enforcement effort. Under Customs' plan, the 701 Forms would be transmitted directly to FDA field offices. The FDA realized that this proposal would put importers on the equivalent of the honor system because Customs would release and then liquidate (grant final approval to) imports without the FDA necessarily knowing that the foodstuffs, medical devices, and other products had been imported. At the request of the Subcommittee, the Customs Service agreed to continue collecting Form 701 for the time being.

In January 1989, Subcommittee staff interviewed FDA and Customs personnel in Seattle, Los Angeles, and San Francisco. Our purpose was to determine if Customs was collecting the Form 701 as it agreed in the December letter.

We confirmed that Customs was continuing to collect the 701 Forms. However, our interviews also revealed the stunning fact that Customs had not been checking to make certain that imports refused admission by the FDA were being destroyed or reexported as the importers claimed. Furthermore, we found that the FDA knew the Customs Service was not checking. The absence of effective Customs policing of imported food rejected by the FDA meant that contaminated foodstuffs were ending up on the dinner tables of American consumers. We have subsequently learned that the FDA knew of the problem in November 1986, or earlier, because they acknowledged it in a staff study of that date. We do not yet know why the FDA and the Customs Service allowed this situation to continue, although in at least one port, the Customs Service cited a shortage of manpower.

The reports of illness because of imported foods filed with FDA demonstrate the seriousness of this situation. During the first five months of 1989, 104 such incidents were reported to the FDA. In 38 of these cases, the illness or reaction was so severe that a physician was contacted. In eleven cases, hospitalization was required. There was one report of a death. FDA provided a file on this death and we would like to offer it as Exhibit II.

In 1988, 149 reports were filed with FDA. In 25 cases, a physician was contacted and hospitalization was required. Such problems are not new. In 1984, one woman died after using herbal pills imported from China and another person required emergency hospitalization. It should be emphasized that these are the cases that came to public attention and were voluntarily reported to the FDA. It is anyone's guess as to how many other incidents went unreported. FDA estimates that one percent or less of the cases were reported. There can be no dispute, however, that Americans have gotten seriously ill from consuming imported foods or drugs and, in a few instances, died.

And it is no wonder, given the steadily rising volume of imports and the constantly shrinking percentage that are inspected by the FDA. The General Accounting Office (GAO), in a report to this Subcommittee dated April 1989, stated that FDA had performed inspections on about nine percent of the 1.5 million imports they reviewed annually, and physically tested about two percent. The two percent analyzed in the laboratory is included in the nine percent figure. Moreover, this figure has shrunk from a level of about twenty percent in the mid-1970s. FDA stated that it achieved the nine percent only by reprogramming resources from other areas.

Aside from the rapidly declining level of testing, the truly frightening aspect of the FDA's sampling program is that the laboratory tests result in a 40 percent failure rate. If the samples tested in the lab bear any relationship to the 98 percent of untested foodstuffs that are imported, a huge volume of contaminated goods are entering the commerce of the United States. . .

The fact that rejected foodstuffs are nonetheless sold in domestic commerce has been known for several years to both the FDA and Customs, but neither agency apparently appreciated the extent of the problem. Customs agents testified before the Subcommittee on April 9, 1987, that a variety of schemes had been employed to evade FDA rejection notices. Containers stuffed with newspaper, items of nominal value, or which were empty, were exported and said to contain the refused merchandise. Alternatively, fake destruction declarations were presented to Customs without even the charade of a container shipment. Yet another tactic was to substitute non-contaminated goods for the rejected items, and ship the acceptable material to another country where it could be sold.

In the near term, more effective coordination between the FDA and Customs is needed, and Customs must perform the necessary follow-up inspection and enforcement to deter violations.

In the long-term, consideration must be given to finding ways to shift effectively the burden onto the importers for assuring the satisfactory quality of their goods.

William L. Schwerner

Under the mechanics of the Federal Food, Drug, and Cosmetic Act, all imported foods and drugs are subject to examination at the time of entry into the United States.

> **FILTHY CONDITIONS**
>
> "The frozen tuna in various degrees of thaw were transported by open truck some 14 miles through the streets of Manila to the plant. The box and product were covered with flies, and caked with deteriorating material.
>
> The receiving area for the tuna, like all areas of the plant, was open, with no air curtains, doors or screens. The product was subjected to birds, insects, and rodents, and once in the receiving area the tuna was placed in filthy dented rusted buckets, and these buckets filled with fish were stacked one upon the other, with the bottom baskets resting directly on the extremely filthy maggot covered floor.
>
> The tuna is held at all times at ambient room temperature. Large fish were eviscerated, and small fish were cooked on the same trays. Both cooked and uncooked tuna were in areas covered with flies, and at risk from fecal contamination by birds roosting, nesting and flying overhead."
>
> Congressman Dennis B. Eckart describing tuna processing in Manila in testimony before the House Committee on Energy and Commerce, September 28, 1989

This does not mean that we examine them all, but they must be presented for examination if we wish. The FDA may administratively refuse a product entry if it only appears to violate the Act. The refusal of entry into the United States of an imported article is subject to judicial review.

By contrast, the FDA does not have an opportunity to examine all domestic products before they are introduced into Commerce. To stop domestic product sales if we choose to, we must initiate court action. Therefore, domestic inspections in some respects, are more important to our enforcement process than are foreign inspections.

I wanted to explain these mechanisms because they tend to balance the type of control the Agency has over foreign and domestic products. As you mentioned, the FDA has finite resources to carry out its legislative mandates.

The Agency has developed planning and management systems to efficiently use these resources in attacking the important consumer protection problems. In allocating field resources, high priority public safety issues are emphasized, but lesser responsibilities are also not overlooked.

Headquarters delegates to its field enforcement managers the flexibility to schedule their work to address the significant problems in their region. The long range planning process, coupled with the shorter range action planning that we have used the last few years, has allowed us to reprioritize resources to address issues such as AIDS, imports, tampering incidents, and most recently the generic drug problem.

I strongly urge the subcommittee to consider FDA's program management system in its total context in evaluating our effectiveness. The suggestion was made some time ago that the agency should perform broad statistical sampling of imports.

I question whether we should divert many of our scarce resources from the current targeting of products most likely to be violative, to such a broad random approach. Indeed, the import product surveys we have done in recent years have revealed very few problems.

This past July, the subcommittee reported a 40 percent violation rate for the imports that we examined. This rate is achieved by sampling after we review the entry document. In other words, a preliminary judgment is made.

I really believe that this rate indicates that our targeting is highly effective when we approach cargo to examine it. Resources are therefore used intelligently to stop many defective products from reaching consumers.

Contrary to some of those recent media reports I mentioned, a review of FDA complaint data for the past 3 years showed no deaths attributed to import canned food. The complaints received do not appear to us to be significantly different than the complaints that we received for domestic products of the same type. . .

I do not wish to belittle the danger inherent in underprocessed foods. However, our import inspectors and consumers also can often, though not always as you said, spot swollen or leaky cans that would be indicative of inadequate processing.

I bring this to your attention because by spotting these problems, they can bring them to our attention through consumer complaints. There is no way for a consumer, on the other hand, to detect a poorly made drug, or a poorly researched one, such as led to the thalidomide tragedy, and the passage of Section 510 of the Federal Food, Drug, and Cosmetic Act.

Accordingly, the public health concerns about foreign produced low-acid canned food, while very real, pale in comparison to the potential drug mishaps that might occur. Most foreign low-acid canned food inspections we have performed over the years were follow-ups to questionable record reviews, import examinations, or for other specified reasons. Whereas, our foreign drug inspections have been scheduled without knowledge of deficiencies. I bring this to your attention only to point out that our management system in the Agency is working to address public health problems.

The transcript of FDA's appropriations committee hearing on March 19, 1987, reveals that Commissioner Young displayed defective canned food, along with other products, and during that hearing emphasized the need to apply more resources to imported foods, and to microbial problems.

The Agency would need many more resources, and foreign governments would no doubt strongly object if the Agency were to attempt inspection of all the manufacturers in the world who ship products to the United States.

I believe consumer protection can be achieved through an active prioritization system, use of our intelligence to target those manufacturers and products that are most likely to be violative, and working with foreign regulatory officials.

Reading and Reasoning

INTERPRETING EDITORIAL CARTOONS

This activity may be used as an individualized study guide for students in libraries and resource centers or as a discussion catalyst in small group and classroom discussions.

Although cartoons are usually humorous, the main intent of most political cartoonists is not to entertain. Cartoons express serious social comment about important issues. Using graphic and visual arts, the cartoonist expresses opinions and attitudes. By employing an entertaining and often light-hearted visual format, cartoonists may have as much or more impact on national and world issues as editorial and syndicated columnists.

Points to Consider

1. Examine the cartoon in this activity.(see next page)

2. How would you describe the message of the cartoon? Try to describe the message in one to three sentences.

3. Do you agree with the message expressed in the cartoon? Why or why not?

4. Does the cartoon support the author's point of view in any of the readings in this publication? If the answer is yes, be specific about which reading or readings and why.

5. Are any of the readings in Chapter Two in basic agreement with the cartoon?

Reprinted with permission of *Star Tribune*.

CHAPTER 3

IRRADIATING FOODS

14. THE BENEFITS OF FOOD IRRADIATION 89
    American Council on Science and Health

15. THE HAZARDS OF FOOD IRRADIATION 95
    National Coalition to Stop Food Irradiation

READING 14
IRRADIATING FOODS

THE BENEFITS OF FOOD IRRADIATION

American Council on Science and Health

The American Council on Science and Health (ACSH) is a national consumer education association directed and advised by a panel of scientists from a variety of disciplines. ACSH is committed to providing consumers with scientifically balanced evaluations of issues relating to food, chemicals, the environment, and health.

Points to Consider:

1. What are the advantages of food irradiation?

2. Why are irradiated foods safe to eat?

3. How have other nations reacted to food irradiation?

4. What is the legal status of food irradiation around the world?

American Council on Science and Health, "Irradiated Foods," December, 1988.

A new technique, irradiation, may soon be an important addition to this arsenal of food protection methods.

Keeping our food supply safe and abundant is the challenging task. Insects and other pests compete with us for it. Microorganisms can spoil it and make it unsafe to eat. Some foods are seasonal and highly perishable; unless safe means of processing and storing them are available, periodic shortages can occur.

Over the centuries, great efforts have been devoted to finding ways to preserve food and protect it from microorganisms, insects, and other pests. Drying was one of the first techniques developed. Fermenting, salting and smoking also have long histories. Later inventions, including freezing, refrigeration, canning, and the use of preservatives and pesticides, have further increased the quality, quantity, and safety of our food supply by protecting it against contamination and spoilage.

A new technique, irradiation, may soon be an important addition to this arsenal of food protection methods. It is already being used to a very limited extent in the United States and is more widely applied overseas. This process, which has been studied by scientists for virtually this entire century, is very promising, yet relatively few American consumers have heard of it. . .

Many foods, including meat, poultry, some types of fish, and some vegetables, are suitable for radiation sterilization, but others are not. For instance, undesirable flavor changes occur in certain irradiated dairy products, so these foods are unlikely candidates for commercial irradiation at the present time. The shelf-stable fluid milk that is popular in Europe and recently introduced in the United States is not an irradiated product. This milk, which can be stored safely at room temperature until it is opened, is preserved by a ultra-high temperature heat treatment, not by irradiation. . .

- **Is irradiation the same thing as cooking in a microwave oven?**

No. Irradiation involves the treatment of food with ionizing radiation. In a microwave oven, foods are exposed to microwaves, a type of non-ionizing radiation that generates heat in moist products.

- **Does irradiation make food radioactive?**

No. The doses and energy levels of radiation approved for the treatment of foods do not have enough energy to induce radioactivity in the food.

- **Does irradiation generate radioactive wastes?**

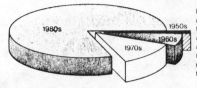

COUNTRY APPROVALS OF IRRADIATED FOODS, 1950–88

Regulatory authorities in 35 countries have approved the use of radiation processing for more than 30 different kinds of foods and food ingredients. Most of these approvals have been granted in the 1980s following the adoption of international safety and health standards for irradiated foods.

Source International Atomic Energy Agency

No. The process simply involves exposing food to a source of radiation. It does not create any new radioactive material. Of course, when radioactive isotopes of cobalt or cesium are used as the radiation source, they must be disposed of properly after they are used. But using these materials for food irradiation does not make their eventual disposal more difficult.

- **Are irradiated foods safe to eat?**

The safety of food irradiation has been systematically and comprehensively evaluated. Extensive studies of low-dose applications of irradiation, such as sprout inhibition and insect disinfestation, have shown that foods treated in these ways are safe to eat. These studies are described in more detail below. Most authorities have deferred final conclusions on the safety of high-dose food irradiation treatments, such as sterilization, until the results of a large American study of radiation-sterilized chicken could be fully analyzed. The evaluation of the results of that study was recently completed, and no evidence of a health hazard was found. This confirms earlier studies indicating that high-dose irradiation is safe. National and international authorities can be expected to take steps toward the approval of high-dose irradiation.

- **How has the safety of irradiated foods been tested?**

Irradiation produces some substances in foods called radiolytic products. "Radiolytic" does not mean radioactive. It simply means that these substances are produced by irradiation. Tests which have been conducted failed to show any of these substances could be harmful. In addition to testing for immediate toxic effects, the tests showed that these radiolytic products did not pose a cancer hazard, interfere with reproduction, cause birth defects, or pose other long-term hazards...

- **Would a food irradiation facility endanger the community where it is located? Could there be a "meltdown"? Would people in the area be exposed to dangerous radiation when radioactive**

> **GLOBAL INTEREST**
>
> *Many countries are interested in food irradiation as a method to reduce sometimes staggeringly high food losses. In Africa, for example, post-harvest food losses are a fundamental cause of food shortages.*
>
> International Atomic Energy Agency News Features, *December, 1988, p. 1*

materials are transported to and from the facility?

A food irradiation plant would not endanger a community, just as the approximately 50 medical products irradiation plants and more than 1000 hospital radiation therapy units now operating in the United States do not endanger the communities in which they are located. Certainly, a food irradiation facility must be constructed and operated properly to ensure that there are no radiation hazards. But this is not a new challenge; the necessary precautions are well understood because they have long been applied in other types of irradiation facilities...

- **Would workers in a food irradiation plant be exposed to hazardous radiation?**

No. As a result of long experience in designing and operating similar types of irradiation facilities, the necessary precautions for worker safety in a food irradiation plant are well understood. These precautions are enforced by several federal agencies in the U.S. The Occupational Safety and Health Administration (OSHA) is responsible for regulating worker protection from all sources of ionizing radiation. Food irradiation plants that use cobalt or cesium as the radiation source must be licensed by the Nuclear Regulatory Commission (NRC) or an appropriate state agency. The NRC is responsible for the safety of workers in NRC-licensed facilities. Plants in the U.S. that use machine-generated radiation fall under the jurisdiction of the FDA, which issues performance standards for such facilities to ensure worker safety.

- **Is food irradiation being promoted because it is a way to dispose of radioactive wastes?**

No. Of the four possible radiation sources for food irradiation, only one — cesium — is a by-product of nuclear fission. This radiation source might therefore be called a waste product, but is rarely described as such, because the term "waste" is usually reserved for substances of no value, and radioactive cesium is very expensive and in short supply. The supply of radioactive cobalt is also limited, and this isotope must be manufactured; it is not a waste product...

- **How have the international health organizations reacted to food irradiation?**

Generally, they have been enthusiastic about its potential uses. After evaluation of the scientific data, they have also become confident of its safety. A 1981 World Health Organization (WHO) document states:

"All the toxicological studies carried out on a large number of irradiated foods, from almost every type of food commodity, have produced no evidence of adverse effects as a result of irradiation..."

- **How have U.S. health and scientific organizations reacted to food irradiation?**

The American Medical Association, in a 1984 letter to a Congressional committee that was considering legislation related to food irradiation, stated that food irradiation is safe, may be an important substitute for pesticides, and can control bacterial contamination in some foods.

The Institute of Food Technologists (IFT), the principal scientific organization in the field of food science, has also evaluated food irradiation. IFT's Expert Panel on Food Safety and Nutrition concluded in 1983 that irradiation was safe and could provide consumers with more food of higher quality. IFT's findings are summarized in the publication "Radiation Preservation of Foods," listed in the "Suggestions for Further Reading" at the end of this report.

- **What is the legal status of food irradiation around the world?**

Over 30 countries have approved some applications of irradiation, and irradiated foods are now marketed in about 20 countries. For instance, in Japan, each year at least 10,000 to 20,000 tons of potatoes are irradiated to prevent sprouting. A wide variety or irradiated foods has been approved in the Netherlands, and plants there are irradiating about 18,000 tons per day of food. About 8,000 to 10,000 tons per year of food is irradiated in Belgium. Two commercial irradiation plants in South Africa process mangoes, papayas, and vegetables. In 1983, as part of a special project, 50 metric tons of onions were irradiated and sold in supermarkets in Chile. Spices and onions are irradiated commercially in Hungary, and they have also been produced in small quantities in France, Israel, Czechoslovakia, and other countries.

- **What is the legal status of food irradiation in the United States?**

In the United States, irradiation has been approved for three specific purposes: maturation delay, insect disinfestation of fresh foods, and control of microorganisms and insects in spices and other seasonings. The first two of these approvals (potato sprout inhibition and insect disinfection of wheat and wheat flour) came in the 1960's, and neither has been taken advantage of commercially. Irradiation of spices was approved in 1983 and is being used commercially to some extent.

Food irradiation is regulated by the FDA under the terms of the 1958 Food Additive Amendment to the Food, Drug and Cosmetic Act. This law prohibits the use of a new food additive until its sponsor has established its safety and FDA has issued regulations specifying conditions of safe use. The law specifically includes "any source of radiation" in the definition of "food additive". . .

- **Now that irradiation is approved, will it be used?**

Most experts predict that it will be used, but the extent will depend on many factors, including regulatory actions, consumer attitudes toward irradiation and toward other processes used for the same purposes, the economics of irradiation and competing processes, and the nature of labeling requirements.

Some people have claimed that there must be little interest in irradiation in the United States, since no one has ever taken advantage of the FDA approval of irradiation for sprout inhibition of white potatoes or insect disinfestation of wheat products. However, good inexpensive alternatives to both of these irradiation treatments have been available, so there has been little economic incentive to irradiate. Furthermore, it may not be economically feasible to operate an irradiation facility for only these two processes, because a food irradiation facility must be utilized for large quantities of food, on a year-round basis, in order to be economically viable. Irradiation is not likely to become common in the United States until FDA approves it for a wide variety of purposes and agricultural commodities. . .

Conclusion

Irradiation could have many benefits for U.S. consumers by increasing the variety of techniques that can be used to provide a safe, wholesome, convenient food supply. Extensive scientific testing has shown that proper use of food irradiation does not present a health hazard. All of the evidence indicates that consumers have nothing to fear from irradiated foods, but instead can look forward to a greater variety of high-quality food products and in certain cases safer products if this process comes into more widespread use in the United States.

READING 15

IRRADIATING FOODS

THE HAZARDS OF FOOD IRRADIATION

National Coalition to Stop Food Irradiation

The National Coalition to Stop Food Irradiation works from its national office in San Francisco, California. They publish a newsletter and other publications in opposition to food irradiation, and represent one of the nation's leading organized groups on this issue.

Points to Consider:

1. How does food irradiation work?

2. Why is irradiated food unsafe to eat?

3. What kinds of food will be irradiated?

4. Will food irradiation damage the environment?

5. Who promotes food irradiation?

Excerpted from a newsletter titled "Food Irradiation," October, 1989, published by the National Coalition To Stop Food Irradiation.

Dr. Barna reviewed 1,223 studies on the wholesomeness of irradiated foods. He found only 185 beneficial effects and 1,414 harmful effects on food and the animals consuming that food.

Food is irradiated to kill insects and bacteria, to prevent sprouting, and to slow ripening. Irradiation works by rearranging the molecular structure of organisms. However, the food itself is also changed by irradiation. Irradiation facilities, using radioactive materials (Cobalt 60 and Cesium 137) or x-ray generators, create serious health and safety hazards and raise grave ecological concerns.

How Does It Work?

Food irradiation takes place in a specially constructed chamber with concrete walls 6-8 feet thick where food is carried past radioactive materials on a conveyor belt. Irradiation levels range from 15,000 rads to kill sprouting enzymes in potatoes to 3 million rads to kill bacteria in spices. By comparison, a chest x-ray gives off about 0.05 rads. The dosage of irradiation given to spices equals *60 million chest x-rays.* When radiation strikes food, it alters and damages cells by rearranging their molecular structure, killing organisms and delaying ripening. But irradiation is an indiscriminate treatment, impairing the food itself: vitamins and enzymes are damaged; new compounds, many of them toxic, are produced. These changes are not detectable so the consumer will be unaware that the food has been irradiated.

Is Irradiated Food Safe to Eat?

We believe that irradiated food is *unhealthy and unsafe to consume.*

- Animals fed irradiated foods have developed testicular tumors, kidney disease, shortened life spans, loss of weight, and increased rate of infertility and death of offspring.
- At currently authorized dosages in the U.S. there is great danger of increased production of carcinogenic by-products in food.
- Cancer research scientist, George Tritsch, Ph.D., says: "The amounts of toxic, radiolytic products formed in irradiated food are irrelevant since food is consumed during the entire lifetime, and only a single carcinogenic insult is needed to produce a malignant tumor."
- There is significant nutrient injury to irradiated food at 100,000 rads, and even greater nutrient destruction at higher dosages. Vitamins C, B1, B2, B6, A, E, and K,

> ### RADIATION ACCIDENT
>
> *What are the stakes involved for us, the potential victims of this industry? Consider that just this last year a Georgia company, Radiation Sterilizers, found a radiation leak. Spot checks were ordered on the products they treated. The result: 70,000 boxes of contact lens solution, medical supplies, and milk cartons had to be destroyed.*
>
> *This time, the contamination was caught because the boxes were still in the warehouse. But what if it had been food, shipped promptly to market and consumed immediately? Is it reasonable for you—for your children—to face the entirely unnecessary risk of food contaminated with radiation just because the nuclear industry needs a market for its wastes?*
>
> Public Letter, National Coalition to Stop Food Irradiation, *1989*

amino and nucleic acids are damaged.

- In the only published study on humans, malnourished children were fed irradiated wheat and developed abnormal blood cells called polyploids, linked to leukemia. The study was immediately suspended and the polyploids disappeared. Similar results were found in studies repeated on rats and monkeys.

- The most extensive survey ever undertaken on irradiated food research was conducted by Dr. Joseph Barna for the Hungarian government in 1979. Dr. Barna reviewed 1,223 studies on the wholesomeness of irradiated foods. He found only 185 beneficial effects and 1,414 harmful effects on food and the animals consuming that food.

- In 1980, using "theoretical calculations in radiation chemistry," the U.S. Food and Drug Administration (FDA) pronounced irradiated food safe and wholesome. However, in the next two years FDA could only find five out of 413 studies it reviewed which "could be said to support safety." Nevertheless, rejecting all contrary evidence, FDA approved its own theoretical hypothesis and continued approving irradiation of various foods.

Where Does the Radiation Come From?

The major sources of gamma irradiation of foods are Cesium 137 and Cobalt 60. Cesium 137 is a by-product of nuclear weapons production and nuclear power generation. Water soluble, it is the most abundant isotope in the nation's nuclear waste and provides treacherous problems for storage.

> **1000 FOOD IRRADIATION FACILITIES**
>
> *Government plans call for the construction of up to 1000 food irradiation facilities across the country—some in the midst of urban neighborhoods; others in rural and farming communities. Irradiation facilities will utilize between one million and ten million curies of radioactive material. In the case of Cesium-137, ten million curies is more than 1000 times the Cesium released by a 20 kiloton nuclear bomb. A 20 kiloton bomb is slightly larger than the bomb dropped on Hiroshima. It is legal to irradiate fruits and vegetables at doses up to 100,000 rads (radiation absorbed doses) while spices can be exposed to three million rads. 100,000 rads is the equivalent of two million chest x-rays.*
>
> Food Irradiation Fact Sheet, *Food and Water Inc., 1989*

Radioactive Cobalt 60 is man-made and its production generates new nuclear waste.

Is It Environmentally Safe?

Since the irradiation service industry (which now deals mostly with medical supplies) began in the mid 1970's, there have been thirteen reported accidents and numerous safety violations.

In June, 1988 an on-going leak was discovered at an irradiation plant in Decatur, Georgia. Contaminated milk cartons and medical supplies were dispersed and only later recalled. This raises the specter of food, irradiated but unknowingly contaminated, being sold and consumed before anyone detects the accident.

One proponent envisions a thousand irradiation facilities operating in the U.S. near agricultural areas, airports, seaports, urban and suburban areas. With the prospect of a thousand irradiation facilities operating in the U.S. and radioactive materials continually transported along highways, the probability of more safety violations and accidents would skyrocket. Radiation accidents irreversibly damage people and the environment.

What Foods Are Irradiated in the U.S.?

By 1980 the Food and Drug Administration had approved the irradiation of wheat, wheat flour, potatoes, and 60 dried herbs and spices and in 1985 authorized the irradiation of pork, followed by fruits and vegetables in 1986.

However, due to strong consumer protest, only some herbs and spices in processed foods are being irradiated currently.

TELL THE WORLD HOW YOU FEEL!

Source: National Coalition to Stop Food Irradiation

FDA is now considering irradiation of poultry.

What States Ban Irradiated Food?

New York and Maine ban food irradiation.

Are Irradiated Foods Labeled?

The FDA does not require labeling of irradiated ingredients. For example, potato soup made with irradiated potatoes need not be so labeled. In fact, millions of pounds of spices are irradiated each year and are not labeled.

The FDA does require that irradiated whole foods carry a written label, but this requirement expired on April 18, 1990, leaving on the label only a flower symbol, the "radura," to indicate irradiation.

Will Food Irradiation Replace Pesticides?

Irradiation proponents claim that food irradiation will reduce dependency on pesticides. Actually, food irradiation is a post-harvest treatment and will not replace one ounce of chemical soil treatment or pesticide used during growth. In all cases irradiation will be used in addition to chemicals applied in the fields. In some cases chemicals will be added after irradiation. Food irradiation has been touted as a replacement for the post-harvest fumigant, ethylene dibromide. Banned in 1983 when it was found to be carcinogenic, very little EDB had actually been used as a fruit and vegetable fumigant. Non-toxic alternatives to EDB, such as CO_2, steam heat, dry heat, vacuum cleaning, and "double-dipping" water baths are cheap, proven, available, healthier and safer than irradiation.

Worldwide Acceptance?

Proponents claim that food irradiation is permitted in 28 countries. However, only limited amounts of a few food items are irradiated commercially. Many countries, including Great Britain, Ethiopia, Denmark, and Sweden, have banned food irradiation altogether. Proponents are pushing food irradiation technology in the Third World where consumers are not well organized and lack information.

Who Promotes Food Irradiation?

The United Nation's International Atomic Energy Agency (IAEA) is vigorously promoting the food irradiation industry worldwide. In the United States its chief proponents are the federal government, also private contractors who stand to profit from widespread acceptance of this technology. Food irradiation research and development is being funded by the U.S. Department of Energy (DOE) with taxpayers' dollars. DOE would like to find some use for its hazardous nuclear waste and is trying to build six demonstration food irradiation facilities in the U.S. (Hawaii, Alaska, Washington, Iowa, Oklahoma, Florida) to promote the food irradiation industry.

Who Is Working For Consumers?

The National Coalition to Stop Food Irradiation has over 90 affiliated groups and supporting organizations around the world, and is working to protect consumer interests across the U.S. To find the affiliate nearest you, write NCSFI.

What You Can Do

- **Educate yourself** on food irradiation and share your concerns. Reproduce and distribute this brochure as often as you like.
- **Avoid irradiated foods.** Tell your grocer that you will not buy irradiated food. Patronize stores that declare a policy to not knowingly sell irradiated foods. You have the right to eat safe food and live in a safe environment.
- **Contact your elected officials.** Urge your Congressperson to co-sponsor and support the Food Irradiation Safety and Labeling Requiring Act (HR 2405 and S 1037).

Over 100 members of Congress have co-sponsored this bill. Write your Representatives/Senators and urge co-sponsorship of HR 2405 and S 1037!

- **Join NCSFI!** Support the work we do and keep yourself informed on this issue of vital concern. Participate in letter and phone campaigns publicized in every issue of *Food Irradiation Alert!*

Reading & Reasoning

Points and Counterpoints: Natural Pesticides vs. Man-Made Pesticides

This activity may be used as an individualized study guide for students in libraries and resource centers or as a discussion catalyst in small group and classroom discussions.

The Point—American Council on Science and Health

The EPA estimated in 1984 on the basis of extensive food testing that the average level of EDB contamination in grain-based foods was 2-3 ppb. . .

It is worth comparing this amount and potency of EDB with several of the natural carcinogens. Aflatoxin B1, is about 1000 times more potent than EDB, yet it is allowed in foods at levels as high as 20 ppb, which is nearly ten times higher than the average level of EDB found, pre-ban, in grain-based food products. . .

Symphytime is about as carcinogenically potent as EDB, but a cup of comfrey tea contains 130 ug of it, or some 260 times as much as a typical day's EDB ingestion. Thus the carcinogenic risk from just this one substance in only one cup of comfrey tea would be equivalent to about 8 months of EDB ingestion at the average pre-ban rate.

These examples could readily be multiplied, but they clearly indicate that nature's pesticides, taking into account both their potency and the amounts in which they appear in food, are substantially more hazardous than the man-made pesticides in food. Clearly, research work needs to be balanced more evenly between natural and man-made substances in food. ("Does Nature Know Best", American Council on Science and Health Pamphlet, April 19, 1989.)

The Counterpoint — Consumers Union of the U.S.

One important subdebate of the "comparative risk" argument involves the relative contribution to dietary cancer risk of naturally-occurring toxic substances in foods, compared with synthetic pesticide residues. This issue has been raised most directly by the work of Bruce Ames, from the Department of Biochemistry at the University of California at Berkeley. . .

In plain terms, the Ames theory is speculation. It is an interesting hypothesis, based on a small number of data points. But the vast majority of information needed to make quantitative comparisons between natural and synthetic carcinogens in foods simply does not exist. Dr. Ames and his followers have extrapolated from a very few data to reach sweeping conclusions, which are opinions, not facts. The argument may seem credible because of Dr. Ames' stature as a scientist at one of the nation's top universities. But in our judgment, it is somewhat reckless to assert that public policy should be based on a hypothesis for which there is as yet so little concrete scientific evidence. . .

A second critical issue is the comparative potency of both natural and synthetic carcinogens in the diet. Some natural carcinogens (such as aflatoxin) are extremely potent. Some of the synthetic pesticides are much less potent carcinogens. But there are also very potent carcinogens among synthetics, and carcinogens of low potency among the natural compounds. . .

The final critical weakness of the Ames argument is in its assessment of exposures. The presence of carcinogens in foods alone is not enough to define the risk; one also needs to know which foods contain it, what the residue levels are, and most important of all, how much people eat of the foods that contain the carcinogen. . .

In summary, when all the weakness in the scientific data on which Ames' estimates rest are examined, the comparative risk information he has generated can be seen for what it is: extraordinarily tentative, uncertain information, subject to extreme uncertainty. (Congressional Testimony by the Consumers Union, November 1, 1989.)

Guidelines

1. Which argument do you agree with the most?
2. Social issues are usually complex, but often problems become oversimplified in political debates and discussions. Usually a polarized version of social conflict does not adequately represent the diversity of views that surround social conflicts.

Examine the counterpoints above. Then write down other possible interpretations of this issue than the two arguments stated in the counterpoints above.

CHAPTER 4

SEAFOOD SAFETY

16. CONSUMING SEAFOOD IS RISKY BUSINESS 106
    Ellen Haas

17. OUR SEAFOOD IS SAFE TO EAT 112
    Lee J. Weddig

READING

16   SEAFOOD SAFETY

CONSUMING SEAFOOD IS RISKY BUSINESS

Ellen Haas

Ellen Haas is Executive Director of Public Voice for Food and Health Policy, a national non-profit consumer research, education and advocacy organization working on issues of health, nutrition and food safety.

Points to Consider:

1. Explain the change in the annual consumption of seafood.

2. Cite specific characteristics which increase seafood potential to endanger health.

3. How is the problem of microbial contamination described?

4. What changes are needed to make seafood safe?

Excerpted from testimony by Ellen Haas before the Subcommittee on Oversight and Investigations of the House Committee on Energy and Commerce, June 5, 1989.

Seafood is the only flesh food in this country not subject to a comprehensive, mandatory federal inspection program.

Risk to Consumers

I appreciate the opportunity to document the serious health risks to consumers in our seafood supply and the inadequacy of current federal and state inspection programs which leave consumers vulnerable to acute illnesses, sometimes death, as well as to risk of cancer.

Since 1983, with our publication of "A Marketbasket of Food Hazards: Critical Gaps in Government Protection," Public Voice has advocated increased mandatory federal monitoring for fish safety as a means of reducing toxic contamination of fish and shellfish. We believe these foods present unnecessary, and too frequently, harmful risks to unsuspecting consumers. . .

American Diet

The Surgeon General and the National Academy of Sciences have warned that the American diet is too high in fat, increasing the risk of heart disease, stroke, and certain cancers. These reports build on previous recommendations from the American Heart Association, American Cancer Society, and the National Institutes of Health. Even the U.S. Department of Agriculture and the Department of Health and Human Services issued Dietary Guidelines for Americans in 1980 that call for diets lower in fat.

Americans have heeded this advice and turned increasingly to fish and shellfish for a low-fat alternative source of protein in their diets. Annual consumption of seafood in the United States increased by over 20% between 1980-1987 and is now well over three billion pounds.

Yet the sad irony of this healthful promise to consumers is that seafood is the only flesh food in this country not subject to a comprehensive, mandatory federal inspection program.

Recently, questions about the safety of seafood have been receiving increasing attention. In January, 1990, a special committee of the National Academy of Sciences, Food, and Nutrition Board began a two-year investigation into the health effects on the consumer from microbial and chemical contamination of fish.

And last year, public concern over fish safety showed up in the marketplace when, after a steady eight-year increase, seafood consumption went down by two percent. The change in public attitudes toward fish is documented in the results of a recently released Harris public opinion poll. While the number of people buying low-fat or low-cholesterol foods rose nine percentage points from the year before, individuals reporting they try to eat fish twice a week dropped six percentage points since 1987.

Illness from Seafood

Seafood has certain characteristics which increase its potential to endanger human health. The risk of acute illness from bacteria and parasites is greater in fish and shellfish, which, unlike most meat and poultry, is generally harvested from uncontrolled environments. The waters from which fish are taken may be polluted with sewage, industrial chemicals, or pesticides.

In addition, certain types of seafood, such as shellfish, actually concentrate contaminants in their tissue. Even when refrigerated, fish and shellfish usually deteriorate quickly. And some seafood is commonly eaten raw.

It is estimated that thousands of Americans become ill each year from eating contaminated fish and shellfish. Disease-causing agents, like ciguatoxin and vibrio cholera bacteria, cause symptoms such as nausea, diarrhea, vomiting, fever and abdominal cramps, among other effects.

A summary of foodborne disease surveillance from the Center for Disease Control (CDC) covering the 12-year period 1973-1984 reported 2,998 foodborne outbreaks of disease with known sources. According to the CDC figures, the total number of reported outbreaks traced to fish and shellfish actually exceeded the total reported outbreaks traced to beef and poultry combined. Seafood accounted for nearly 10% of the outbreaks, beef represented nearly 5%, and poultry 3.8% of the total outbreaks with a known source.

This figure takes on added significance because we eat more than four times as much beef and three times as much poultry as we do seafood.

When examined together with USDA consumption data for each of these food groups for the same period, the data show that the relative risk of illness outbreaks from seafood compared to other meats is very high. In fact, the risk of a food poisoning outbreak from eating seafood is actually 10.4 times greater from seafood than it is for beef, 6.9 times greater than for poultry, and 5.8 times greater than for pork. . .

At FDA's December 16, 1988, National Meeting on Cooked/Processed Seafood, Dr. John E. Kvenberg of FDA's Center for Food Safety and Applied Nutrition reported on the 1988 microbial testing of processed seafood. Because processed seafood will usually not be cooked again before it's eaten, the presence of pathogens causes special concern. 14.4 percent of the 369 domestic samples of processed seafood were rated Class 3—enough pathogens were present to enable FDA

> ## HAZARDOUS, IF NOT DEADLY
>
> *The trouble is, some of the fish coming into United States supermarkets and restaurants also can be hazardous to your health, if not deadly. Seafood can carry harmful bacteria, viruses, natural toxins and parasites. Residues from untreated sewage, industrial waste and pesticides flow into lakes, rivers and oceans, and accumulate in fish and shellfish.*
>
> Jay Stuller, American Legion Magazine, *April, 1989, p. 23*

to take official action, (e.g., a recall.) Crab meat was contaminated frequently; 28.8 percent of the 165 samples were rated Class 3; 10.2 percent of the domestic 118 cooked shrimp samples also were actionable. Overall, the imports fated slightly better with 11.1 percent of the samples falling in the Class 3 category. The types of pathogens found included Listeria monocytogenes, vibrio cholera and E. coli.

At the same FDA meeting, Richard Ronk, Acting Director of FDA's Center for Food Safety and Applied Nutrition explained that certain seafood companies do not follow good manufacturing practices, which exacerbates contamination problems. He said: "This problem of microbial contamination must be brought under control. It is widely recognized that the microorganisms readily adapt to changing environments. . .It is well documented that certain segments of the seafood industry are not conforming to current good manufacturing practices in handling their products. Cross-contamination of raw and processed products and improper storage temperatures are examples of the problems mentioned today which need to be addressed by the seafood industry."

Unfortunately, these industry practices are not unique. Some include:

- Trash dumpsters without covers located outside a plant on its landing area. Seagulls were constantly attracted to the garbage. Their excrement littered the area on which seafood was delivered in open crates.
- Wooden crates and totes which cannot be properly sanitized were used without plastic liners to hold raw and then finished cooked product.
- Workers entering a plant walked through raw seafood juices before entering processing area and the firm did not provide a foot bath to cleanse their shoes.

Contamination of Seafood from Polluted Waterways

Pesticides and other chemical compounds which are present in waterways because of industrial and municipal dumping, agricultural run-off, and application of pesticides to marshes and shorelines pose a very real hazard of public health.

These chemical residues resist degradation, are incorporated into phytoplankton which is the first link in the food chain, and become concentrated in the tissues of fish and shellfish consumed by humans. Many of these residues present long term health risks to consumers. Some have been linked to cancer and other adverse health effects.

Even the limited data available on chemical contamination of fish demonstrates that toxins are commonplace in the seafood we eat. Of all the domestic food samples collected and analyzed by FDA between 1979 and 1985, including vegetables and dairy products, fish had the highest percent of samples containing illegal residues, 5.2 percent, whereas the market basket average was 2.9 percent...

Mandatory Federal Inspection Needed

American consumers need protection from the health risks associated with eating fish and shellfish and they need it now. Public Voice, along with a national coalition of 35 consumer, environmental and health organizations, has called for a comprehensive, well-integrated federal seafood inspection program. We recommend the following elements for inclusion in such a program:

Mandate certification of fishing vessels to meet sanitation requirements such as proper icing facilities, with periodic unannounced spot checks to ensure proper maintenance.

1. Mandate good manufacturing practices and establish unannounced spot checks of fish processing facilities, several times a year.

2. Establish federal microbiological and chemical standards with meaningful residue sampling programs.

3. Improve import programs for contaminants.

4. Establish a traceback and record keeping system for fish and especially shellfish of their source or origin.

5. Increase enforcement activities for federal officials and strenghten enforcement authority by means such as civil penalties.

6. Appropriate funding for development of needed test methodology for toxins.

7. Establish a program to address economic fraud (deceptive labeling of products, glazing).

8. Establish improved consumer information programs that include educating physicians about risks, especially of eating raw shellfish, to immune-suppressed people.

Many Americans are still unaware of the potential dangers of seafood, and they are turning to fish as one way of eating a healthier diet. Unless legislation establising a comprehensive, mandatory seafood inspection program is enacted and signed in this Congress, those same health conscious consumers will be putting themselves at increasing risk of serious illness, even death.

READING 17

SEAFOOD SAFETY

OUR SEAFOOD IS SAFE TO EAT

Lee J. Weddig

Lee J. Weddig is the Executive Vice President of the National Fisheries Institute. He made the following comments about how the commercial seafood industry has kept fish safe to eat.

Points to Consider:

1. Why will seafood consumption continue to grow?

2. Explain the relationship of seafood inspection to the national economy.

3. How well is the seafood industry regulated?

4. What changes are needed in the national seafood inspection system?

Excerpted from testimony by Lee J. Weddig before the Subcommittee on Fisheries and Wildlife Conservation and the Environment of the House Committee on Merchant Marine and Fisheries, June 7, 1989.

Seafood products which make up virtually all of consumption exhibit an excellent record of safety.

Twenty-two years ago, the late Senator Philip Hart spoke at the annual convention of the National Fisheries Institute and outlined his belief that there was need for a better seafood inspection program. Senator Hart subsequently introduced legislation which was investigated in hearings, actually passed by the Senate, but did not make it into law.

At the time, the seafood industry was struggling. The Soviet and other foreign fishing fleets were decimating our resources; the Catholic Church had just changed its dietary rules, allowing meat to be eaten on Fridays; and a well publicized incident involving botulism in smoked Michigan whitefish had scared consumers.

A lot has changed since 1967. The Magnuson Fisheries Conservation and Management Act provided means in 1976 to control foreign and domestic fishing pressure on the resources. Fish has become an everyday, not a Friday, food, and the previously unknown potential of botulism in smoked fish is safeguarded by processing and handling procedures.

The most dramatic change since 1967 has been the growth of the fish and shellfish in the American diet. In 1967, per capita consumption was 10.6 pounds edible weight. Last year it was 15 pounds, an increase of more than 46 percent. Even more significant is the growth of fresh and frozen products, as opposed to canned. Here the increase was from 5.8 to 9.6 pounds, a 65 percent increase.

Growth in Consumption

For most consumers, fish and shellfish are no longer an occasional food, or special treat. Seafood is a full fledged every day choice of consumers, along with beef, pork and poultry. The public realizes that fish provide excellent nutrition. . .low fat, low cholesterol and easily digested proteins.

These characteristics, together with the unique tastes and flavors of properly prepared seafood, and the desire for variety, strongly suggest that seafood consumption will continue to grow.

Such growth is good for the country. Health experts advise that heart disease can be alleviated by dietary changes which include more fish. Improvements in public health have direct benefits for the country. Certainly, quality of life is improved with variety; and the nation's economy benefits from a vibrant primary industry which creates income and jobs from renewable natural resources.

These benefits will accrue in the future only if consumers have full confidence in the safety, wholesomeness, and integrity of fish and seafood products. The National Fisheries Institute believes this confidence is in jeopardy and will be maintained and strengthened only through an improved regulatory system which includes mandatory inspection...

The Economy

Bringing the issue of seafood inspection to a successful conclusion is important to the country since the impact on the economy of the fish and seafood industry is far more extensive than most people realize. A recent study conducted by the National Fisheries Education and Research Foundation shows that direct, indirect and induced economic impacts attributable to fish and seafood exceed $80 billion annually. More than one million full-time equivalent jobs are generated. These are big numbers and reflect the myriad pathways on which fish products move from the dock to the dinner table. The study shows that consumers pay about $30 billion annually for fish and seafood with about two-thirds being consumed in restaurants and other food service establishments.

While many of the economic impacts occur in the metropolitan areas, since that is where the value of distribution and service is realized, the primary economic impacts are generated in rural areas where jobs are needed.

The seafood industry is very trade dependent, for both our markets and supplies. More than 3 billion pounds of product worth $5.5 billion were imported last year. This was slightly greater than imports of meat and poultry in volume, but substantially higher in value. As for exports, fishery products soared in 1988, exceeding one billion pounds and $2.1 billion, making the U.S. the world's largest fish exporter. The value of fish exports exceeds that of meat and poultry.

It is evident that seafood is a major component of our nation's food supply—a component important enough to the consumer and the economy that our ability to maintain the confidence of consumers here and overseas must be protected for the common good.

Popular Opinion

Contrary to popular opinion, the fish and seafood industry is not without regulation. We are subject to the Food, Drug and Cosmetic Act and the Fair Packaging and Labeling Act. These laws are enforced by the Food and Drug Administration. Enforcement includes plant and product inspections and port-of-entry inspection for imports. The extent of this enforcement activity varies, but it does exist and has been documented in such reports as the General Accounting Office study of August, 1988 on Seafood Safety and the responses of

> **PROGRESS**
>
> *Fortunately, progress is being made through another avenue in the meantime. The U.S. Department of Commerce's National Marine Fisheries Service, together with the supermarket industry's Food Marketing Institute, has just kicked into place a voluntary Standard of Excellence Award Program to encourage ongoing seafood inspection and high standards of fish safety in grocery stores.*
>
> Tufts University Diet and Nutrition Newsletter, *September, 1989*

the FDA to questions raised by the Chairman of the House Subcommittee on Oversight and Investigations of the Energy and Commerce Committee.

Various states also enforce their laws relating to food safety. In respect to molluscan shellfish, the coastal state governments, in coordination with the FDA, operate the National Shellfish Sanitation Program. Finally, the U.S. Department of Commerce provides a for-fee voluntary inspection and grading service under the Agricultural Marketing Act of 1946.

The actions of overseas governments also have a bearing on seafood safety. Such major supplier nations as Canada, Norway and Iceland, among others, have seafood inspection programs specifically designed for fish and seafood.

Present System Inconsistent

The problem is not that there is no inspection system, but rather that the present system is inconsistent and incomplete and thus does not provide the level of assurance demanded by the public. This demand is strong and it's becoming evidenced in direct fashion. Certain consumers are reducing or avoiding fish consumption. Recent surveys by the Food Marketing Institute showed 15 percent of consumers are avoiding fish for safety reasons. A large part of the current demand that more be done to assure the safety of fish stems from concern over the marine environment which has suffered from decades of toxic substance discharge, substandard sewage treatment, oceanic sludge dumping, pesticide run-off, and oil spills. While there is little evidence to show that present levels of toxic chemicals in fish have actually caused illness, there remains a nagging concern over the long-term impacts.

Illness and Seafood

Examination of the data on actual illnesses and deaths attributable to seafood, and analysis of the data's meaning, do not show widespread problems. The GAO report and documentation by the National Marine Fisheries Service and the FDA attest to this. A larger study now in progress at the National Academy of Sciences can be expected to shed more light on the extent of real problems attributed to seafood as well as to offer recommendations on how these are to be corrected.

It does appear that the major sources of illness attributable to seafood are those caused by contaminated molluscan shellfish when eaten raw; by ciguatoxic reef fish, a problem confined primarily to our tropical island territories; and scombrotoxic fish, a problem unique to a certain class of fish in which histamine can build up if the product is mishandled.

Seafood Products Are Safe

It is important to note that the seafood products which make up virtually all of consumption exhibit an excellent record of safety. It is equally important to note that many of the actual cases are the result of natural environmental conditions, the ciguatoxic fish and some of the raw mulluscan shellfish to be specific. Also, although many fish are caught recreationally, the impact of sports-caught fish in the illness statistics is not known.

While the degree of risk of illness attributable to seafood can be, and is, debated, NFI believes that the regulatory system should be intensified in all areas in which any type of statistical evidence of illness exists. Certainly, a new regulatory scheme must specifically address the problems of ciguatoxin, scombrotoxin and the unique natural toxins, bacteria and viruses at times found in mullusks.

New System Needed

We do not wish to minimize or discount the problems currently affecting relatively small percentages of our products. However, our basic position is that a new system is needed to anticipate rather than react to problems. Such a program is needed because of significant changes occurring in our industry.

There are three federal agencies which could be given the seafood inspection task—the Food and Drug Administration; the Department of Commerce; and the Department of Agriculture. Each of these agencies offers strengths and each has some weaknesses.

Role of FDA

The Food and Drug Administration already has much of the legal authority needed. It has impressive credentials and expertise. It already inspects foods at the ports of entry and administers the National Shellfish Sanitation program. It also works closely with state health agencies. . .

More Regulation

The industry recognizes that the move to more intense regulation will not be simple or painless. But, the record shows that the U.S. fish and seafood industry has matured greatly since the time Senator Hart first broached the subject of inspection more than 20 years ago.

The very introduction of legislation in the past caused a re-examination of practices in the industry with resulting improvements in quality and wholesomeness. These improvements have contributed to the growth the industry has achieved.

Reading and Reasoning

INTERPRETING EDITORIAL CARTOONS

This activity may be used as an individualized study guide for students in libraries and resource centers or as a discussion catalyst in small group and classroom discussions.

Although cartoons are usually humorous, the main intent of most political cartoonists is not to entertain. Cartoons express serious social comment about important issues. Using graphic and visual arts, the cartoonist expresses opinions and attitudes. By employing an entertaining and often light-hearted visual format, cartoonists may have as much or more impact on national and world issues as editorial and syndicated columnists.

Points to Consider

1. Examine the cartoon on page 76.

2. How would you describe the message of the cartoon? Try to describe the message in one to three sentences.

3. Do you agree with the message expressed in the cartoon? Why or why not?

4. Does the cartoon support the author's point of view in any of the readings in this publication? If the answer is yes, be specific about which reading or readings and why.

5. Are any of the readings in Chapter Four in basic agreement with the cartoon?

CHAPTER 5

ORGANIC FARMING AND BIOTECHNOLOGY

18. ORGANIC FOODS: AN INTRODUCTION   120
    Ginia Bellafante

19. THE CASE FOR ORGANIC FOODS   127
    Colman McCarthy

20. THE CASE AGAINST ORGANIC FOODS   131
    Warren T. Brookes

21. BOTH ORGANIC FARMING AND PESTICIDES ARE NEEDED   135
    John Hood

22. BIOTECHNOLOGY AND AGRICULTURE: THE POINT   141
    W.P. Norton

23. BIOTECHNOLOGY AND AGRICULTURE: THE COUNTERPOINT   148
    John Nicholson

READING

18   ORGANIC FARMING AND BIOTECHNOLOGY

ORGANIC FOODS: AN INTRODUCTION

Ginia Bellafante

Ginia Bellafante wrote the following comments in Garbage *magazine. In preparation for the article, the author attended an organic farming conference held by the Natural Organic Farmers Association in New England.*

Points to Consider:

1. How does one define the word "organic"?

2. What is the best way to shop for "organic" foods?

3. How do states regulate the status of "organic" foods?

4. What does the term "certified organic" mean?

Ginia Bellafante, "Organic Foods: Are You Getting What You Pay For?". Excerpted with permission from the Nov/Dec 1989 issues of *Garbage Magazine,* Brooklyn, N.Y., ©1989

No longer a counter-culture occupation, organic farming—the practice of growing fruits, vegetables, and grains without chemically synthesized pesticides, fertilizers, fungicides, herbicides, or post-harvest preservatives—has become a respected science. Universities offer degree programs in environmentally sound farming techniques. Chemical companies including DuPont and Monsanto are researching ways that bioengineering can be used to deter pests naturally.

Organic farming is also becoming big business. Huge growers such as Dole and Sunkist have begun to set aside land for food making its way onto the shelves of major supermarket chains—and into the dishes of some top chefs, who insist that organic greens taste far better than those grown conventionally. On the average, fruits and vegetables grown without pesticides cost 15 to 30 percent more than ordinary produce...

The Wall Street Journal has predicted a nine-fold growth in organic food production over the next ten years. In the absence of an enforced regulatory system, the expansion of the organics market is likely to be accompanied by instances of fraud.

The Panic for Organic

The daily papers are full of reports warning us that some pesticides have been linked to cancer and birth defects, and that others may be damaging to the immune, nervous, and reproductive systems. Of the 600 pesticides registered with the U.S. Environmental Protection Agency (EPA), 496 are known to leave residues on food. In October 1988, the EPA reported that there is at least "limited evidence" that 66 of the pesticides sold for use on food crops may be carcinogenic...

What Organic Means

Under the circumstances, it's not surprising that people are clamoring for organic foods. But be aware that there is no national definition of organic. Some states regulate the organic food industry, but the federal government does not.

There are respected, reputable certifiers at work in states that don't regulate organic farming. But we should be wary and aware of the requirements a particular certifier imposes on its growers. There are approximately 40 organizations in the United States right now that scrutinize the farming methods of growers who want the credibility that comes from being "certified organic." Most of these organizations certify foods exclusively in their own region. Farmers apply for certification voluntarily and generally pay about $150 to do so. The cost of mandated soil tests is additional. Certification committees are usually made up of about 50 percent farmers, and 50 percent consumers and agricultural specialists.

The guidelines issued by various certifiers differ slightly. Many

Illustration by Carol*Simpson.

specify that a certified organic farmer be growing on soil that has not been treated with any chemicals for three years. Others demand that soil be untreated with pesticides for three years, but free of chemical fertilizers for only one or two years. Certifiers usually govern not only what farmers can't use, but also what they can and should use to produce healthy crops. These organizations generally instruct farmers to deter pests by rotating crops and using botanically derived pest controls and natural predators; to keep weeds in check by planting cover crops; and to fertilize soil using only "natural organic" fertilizers like blood and bone meal, and composts made from manure and food and yard wastes. Nutrient-enriching cover crops and compost are vital for healthy erosion-resistant soil as well as healthy plants. . .

The Future of Organic Foods

A surefire guarantee that an item of produce marked "certified organic" is 100 percent chemical-free may never be possible. But there are ways to ensure that the organic produce you're buying is reasonably pesticide-free—and fairly priced. State-by-state definitions of organic not only make it difficult for us to know what we're getting, but they also create enormous obstacles for those trying to market organic produce across state lines. This in turn keeps supplies low and prices high.

A national definition of organic, one that would stipulate for how long a farm must be chemical-free, could eliminate much of this confusion. Senator Wyke Fowler of Georgia has introduced an amendment to the 1990 farm bill that would require that a farm be chemical-free for three years to qualify as organic. It also calls for the formation of an Organic Food Commission "to determine the feasibility and advisability of establishing a national program for the certification of organic food." Indiana Senator Richard Lugar has also sponsored a bill that calls for a reduction of chemical use on farms; but the bill doesn't specifically address the issue of organic farming. . .

How to Shop Organic

Until organic foods become less costly (which is likely as large distributors begin to see the profits to be made in shipping them), and until they are more uniformly regulated, shoppers need to decide whether or not these fruits and vegetables are worth all the extra cash. Here are some tips to keep in mind:

- Always stick with food labelled "certified organic". In some states, there is no definition of organic and there are no certifiers, state or private. Unless you buy certified organic food shipped from a reputable out-of-state certifier, the chances are high you'd be wasting your money on "organic" food just labelled organic.

- In states that have legislated a definition of organic but do not themselves certify or contract out to an independent certification group, inspection of organic food may be minimal or non-existent. On the other hand, there may be one or more extremely reputable private certifiers in the area. If you live in one of these states, you must learn about the requirements those certifiers impose on organic farmers.

- All produce labelled "certified organic" should include the name of the certifying body. You may decide that food certified by an organization that does not require farms to be free of pesticides and chemical fertilizers for three years simply isn't worth the premium price. Send a self-addressed stamped envelope to the Organic Foods Production Association of North America, P.O. Box 1, Belchertown, MA 01007 for a complete list of certifiers across the country. Certifiers should send you their guidelines upon request.

- Buy directly from farmers when you can. Because organic produce contains no preservatives, it may not be fresh by the time it gets to the supermarket. If you find a farmer who claims to be growing organic produce but isn't certified, ask him how he controls pests and weeds. And ask him how he fertilizes his soil.

Status of Organic: State By State

States that define organic for labelling purposes, i.e. organic has a legal meaning but the states don't inspect organic farms to see that definition is adhered to : California, Connecticut, Iowa, Maine, Massachusetts, Montana, Nebraska, North Dakota, Oregon, South Dakota.

States that define organic and contract out to independent certifiers who do the actual certification: Minnesota, Ohio.

States that define organic and do the actual certification: Colorado, New Hampshire, Oklahoma, Texas, Washington.

There are four distinct ways a state may be involved in supervising organics: (1) A state may legislate a definition of organic and certify or inspect farmers who grow organically. (2) A state may define "organic" legally, but contract out inspection to an independent certification organization. (3) A state may merely legislate a definition of organic for labeling purposes but may not check to see that food labelled organic is grown in accordance with that definition. Independent certifiers may operate in these states, however. (4) A state may have absolutely no legislation on organic growing. Independent certifiers do exist in many—but not all—of these states.

Seventeen states have regulations or laws pertaining to organic food. However, legislation differs considerably from state to state. Consequently, an organic apple grown in Massachusetts may not qualify as such in Texas. A California statute, for example, permits food to be labelled organic if it has been grown on soil treated without chemicals for one year. A Texas law requires that organic food come from farmland that has been pesticide-free for three years—the time it usually takes for residues to disappear—but free of chemical fertilizers for only one year.

How these states ensure that growers are complying with state standards is another thorny issue. Only Texas, Washington, Colorado, Oklahoma, and New Hampshire actually certify that organic merchandise has been grown in accordance with state-approved standards. Judgments as to who may be certified are based on comprehensive examination of a grower's practices and a series of soil tests. The Texas Department of Agriculture not only enforces its standards, but also aggressively markets the organic produce it certifies. According to Keith Jones at the Texas Department of Agriculture, 10 percent of the produce in the state's major chains, HEB, Tom Thumb, and Apple Tree, is certified by the state.

A Minnesota statute requires that produce labelled "certified organic" be certified as such by the Organic Growers and Buyers Association, a private, non-profit certifying body based in that state. Minnesota was the first state to form a cooperative relationship with an independent certifier. Other states have begun to follow suit.

The Other Organics

Finding reputable sources of organically grown meats, dairy products, breads, cereals, etc., is even more of a challenge than finding organic produce. Very few certifiers have devised standards for these foods.

The few organizations that have set guidelines for livestock require that certified organic meat and poultry come from animals that have been fed only organically grown grains, and have never been treated with synthetic growth hormones or "subtherapeutic" antibiotics (drugs administered to prevent disease).

Some researchers suspect that growth hormones leave people vulnerable to strains of Salmonella that are not easily curable. The European market became so concerned about the effects of growth hormones on human beings that it stopped buying American meat treated with those hormones in January, 1990. Some scientists also claim that the antibiotics used by ranchers may cause cancer.

Certifiers who deal with livestock claim that few ranchers can meet their standards, in part because supplies of organic feed are limited. As a result, the amount of certified organic meat and poultry available is small. Certified organic milk and eggs must come from certified organic livestock, and as a consequence supplies of organic dairy products are just as slim.

Breads, pastas, and other processed-grain foods labelled organic should contain only ingredients that have been grown organically. These products should also be free of any chemical preservatives or processed additives. The Organic Crop Improvement Association (OCIA) is the most respected nationwide certifier of processed foods that manufacturers claim are organic. Steer clear of products that do not bear the OCIA label because they may not be worth the additional cost.

A Buyer's Guide to Organic Terms

It's easy to get lost in the thicket of technical terms used by organic growers and certifiers. To make shoppers' lives easier, we put together the following guide to organic terminology:

BIODYNAMIC/BIOINTENSIVE: A method of organic farming based on the writings and lectures of Austrian philosopher Rudolph Steiner and biochemist Ehrenfried Pfeiffer. The two men recommend the use of highly complicated and often peculiar compost recipes. Followers of the biodynamic method may be certified organic, but not necessarily.

CERTIFIED ORGANIC: Certified organic farms have been carefully inspected by either a representative of a state department of agriculture or a private certification group to verify that a grower has used absolutely no chemicals on his farm for a specified period of time (usually one, two, or three years). This is your best guarantee that you're getting what you paid for.

INTEGRATED PEST MANAGEMENT (IPM): IPM crops are grown using a combination of biological and chemical methods, with the emphasis on reducing the amount of synthetic pesticides used.

NATURAL: A marketing gimmick used frequently on processed foods, which means absolutely nothing.

PESTICIDE-FREE/CHEMICAL-FREE: Buyer beware! This term does not mean produce has been grown organically. The only way to be sure that food is reasonably free of pesticides is to buy those items labelled "certified organic".

TRANSITIONAL ORGANIC: This phrase implies that the grower has switched from chemically intensive farming to organic methods but must wait one to three years, depending on the rules of his certifier or his state, before his produce can be labelled organic or certified organic. Most chemical residues disappear over a one to three year period.

UNSPRAYED: Unsprayed means that the edible parts of a crop have not been treated with pesticides, although synthetic fertilizers and fungicides may have been used in the soil.

READING

19 ORGANIC FARMING AND BIOTECHNOLOGY

THE CASE FOR ORGANIC FOODS

Colman McCarthy

Colman McCarthy is a political columnist for the Washington Post. *His articles on national and international issues are nationally syndicated.*

Points to Consider:

1. Describe the extent of organic food sales nationwide.
2. What is the major myth about organic farmers?
3. How does the author refute the notion that organic food is expensive?
4. How has pesticide use affected crop loss from insects?
5. Who is Jim Hightower and what did he say?

Colman McCarthy, "The Turn Toward Organic Food," *Minneapolis Star and Tribune*, March 25, 1989. ©1989, Washington Post Writers Group. Reprinted with permission..

In a food industry survey last year, 75 percent of shoppers listed pesticide contamination as their greatest concern.

Before Alar and apples, and before Meryl Streep came to Congress to talk of pesticides in food, I could get in and out of the organic food section of my local co-op in a few minutes. No more. Organic is in. There's no time to linger and squeeze the squash. Grab it and shuffle on.

Organic Sales

Exiles from Safeway now want a safer way, beginning with fruits and vegetables grown with God's help, not the chemist's. Recently, sales of organic food have surged. Even without bitter apples and sour grapes, a national awakening was under way.

In a food industry survey last year, 75 percent of shoppers listed pesticide contamination as their greatest concern. Whole Foods Market, the largest retailer of organic food in Texas, had $61 million in sales last year, up 14 percent from 1987. Nationally, organic food sales were estimated at $3 billion last year. Demand is expected to double within five years, according to the Organic Foods Production Association of North America.

Vegetarian Times magazine has 170,000 subscribers, double the number three years ago. "The Necessary Catalogue" published in New Castle, Va., for those wanting to grow food in a "safe, gentle way avoiding the harsh chemical approach on conventional agriculture," has a subscription list of 10,000 commercial farmers. It was 1,500 five years ago.

Major Myth

As this tributary becomes the mainstream, myths are falling like ripe Alar-free apples. A major myth is that a typical organic farmer is a mantra-humming hippie growing bean sprouts in the back yard with seaweed as mulch and a fly swatter for a pesticide. Those aren't the constituents of Jim Hightower, the Texas agriculture commissioner. No hippies or seaweed-mulchers are among the 60 farmers who enrolled in the Texas organic certification program, the most comprehensive in the nation. Hundreds more applications are pending, even though the program is less than a year old.

Hightower, in Washington last week for the National Conference on Organic and Sustainable Agriculture, talks about fellow Texans who have gone organic: "We've got a citrus producer in the Lower Rio Grande Valley, one of the most chemically intensive farming areas of our state, growing 70 acres of oranges and grapefruit organically. Last month, he shipped

"They want to know if it's been tested for EDB contamination"

Illustration by Carol*Simpson.

his first full truckloads to both the east and west coast...A farmer in Childress, in north Texas, is now selling Texas Department of Agriculture-certified organic watermelons and cantaloupes nationwide...A rice farmer near Beaumont is now selling organic rice through a company in Minnesota but is working to develop his own label. We recently certified the 308 acre organic farm of the Center of the Retarded outside Houston."

Second Myth

A second myth is that organic food is expensive. Joseph Dunsmoor of Organic Farms, a Beltsville, Md., distributor, says that, instead of getting overheated about the prices of organic food, we ought to be adding in the hidden costs of pesticide food: "Billions are paid—not at the cash register but in taxes and bills—to cover the Superfund and cleaning up the environment and health losses due to cancer and toxic diseases."

Dunsmoor believes that the economics of U.S. farming are changing. After World War II, agriculture went west where water, land, energy and labor were cheap. Now it's expensive. A way out is to regionalize production and distribution. "For easterners, bring agriculture back east," Dunsmoor argues. Small is beautiful; so is regional.

> **EBDC FUNGICIDES**
>
> *The Environmental Protection Agency (EPA) called Monday for curbs on the pesticide EBDC, saying its widespread agricultural use poses an unreasonable cancer risk.*
>
> *The EPA proposed eliminating the use of the EBDC family of fungicides on 45 crops, but said its continued use on another 10 food products, including grapes, onions and cranberries, does not present an unreasonable risk to consumers.*
>
> Associated Press, Star Tribune, December 5, 1989

Third Myth

A third myth is that if we don't spray with pesticides, bugs will overrun farms and apples will be wormy. Any organic farmer, or even the mildly alert weekend shopper at the local co-op, knows the falsity of that. Centuries before U.S. chemical companies presented themselves as saviors of the American stomach, beneficial insects served as natural predators. With the heavy introduction of pesticides in the early 1950s, nature's ecological balance was skewed.

Citing overall pesticide production at 2.7 billion pounds in 1983—up from 200,000 pounds in 1950—the National Coalition Against the Misuse of Pesticides reports that during this enormous increase "crop loss due to insect resistance has doubled." A vicious circle has become a vicious triangle: more pesticides, tougher bugs and less food.

Stewards of the Land

Organic farmers are as skilled in the use of biological, botanical and mechanical controls of insects as conventional farmers are in spraying chemicals. They are stewards of the land, not exploiters. Even if a rare worm does wiggle into an apple, no one gets cancer from it. Nor from a splotch on an orange, nor from a carrot that isn't perfectly shaped.

Ten years ago, people who made food choices based on health, nature and ethics were still being ridiculed as flakes, while those with poisoned-based diets were seen as normal. The tables—ladened with organic foods—have turned.

READING 20

ORGANIC FARMING AND BIOTECHNOLOGY

THE CASE AGAINST ORGANIC FOOD

Warren T. Brookes

Warren T. Brookes is a national syndicated columnist. His articles on political affairs appear on a regular basis in Human Events *and the* Conservative Chronicle.

Points to Consider:

1. Why may organic foods be more dangerous than foods raised using pesticides?

2. Who is Dr. Bruce Ames and what did he do?

3. What are natural pesticides in food and how dangerous are they to eat?

4. How dangerous are man-made pesticides?

5. Why are Americans healthier today?

Warren T. Brookes, "Why Organic Foods Are Not Safer for You," *Conservative Chronicle,* August 30, 1989. By permission of Warren T. Brookes and Creators Syndicate.

Organic foods are selling because Americans have been sold unreasonable fears of modern technology.

A gigantic and costly hoax is about to be perpetrated on American consumers in the name of safety and health.

The hoax is a plan supported by the fruit and vegetable industry (who see a premium price bonanza) to get Congress to adopt the first-ever national standards for labeling "organic foods" so consumers can be sure they are "safe".

Price and Safety

Yet foods raised "organically," that is, without using chemical pesticides, are not only 25 percent to 50 percent higher in price, they are also potentially more dangerous to your health than those using conventional pesticide agriculture.

That is the clear implication of the work of one of the nation's top cancer researchers, Dr. Bruce Ames, chairman of the Biochemistry Department at the University of California at Berkeley.

Ames was once the hero of the environmental (and organic foods) movement, because he developed the well-known Ames test used to expose the dangers of pesticide carcinogens.

But over the last decade or so of further research into food chemistry, he has discovered two important things that made him conclude that chemical pesticides at normal levels are "harmless" and foods raised without them might even be more carcinogenic.

Natural Pesticides

In a recent interview he told us: "First, 99.9 percent of all the pesticides we ingest, by weight, are natural, produced by the fruit and vegetable plants themselves as part of their protective mechanism. Probably 5 percent of every plant's dry weight is in natural pesticides.

"Importantly, the proportion of natural pesticides that test positive on rats as carcinogens (about half) is about as high as for synthetics. If this trend continues, it suggests that we are ingesting up to 10,000 times the weight of natural carcinogens in fruits and vegetables as of synthetics."

But Ames hastens to add, "This doesn't mean fruits and vegetables are dangerous. In fact, as last March's National Research Council report demonstrates, eating lots of fruits and vegetables is a good way to prevent cancer and promote health. I am just trying to help you understand that at such low doses, carcinogens—natural or man-made—are much less dangerous than we thought."

Source: Natural Resources Defense Council

Pesticide Dangers

Indeed, it was Ames' discovery of the real extent of our ingestion of "natural carcinogens" that made him realize the tiny traces of synthetic pesticides were probably meaningless.

For example, he points out, "everyone worries about minute amounts of dioxin, but there is lot more of a dioxin-like compound naturally in broccoli than you will ever be exposed to through dioxin contamination in the environment."

Ironically, Ames and others have discovered that fruits and vegetables cultivated to be organically-resistant to pests can have even higher levels of natural pesticide carcinogens:

"You have to understand that plant evolution is a process of continual chemical warfare, with plants constantly generating chemical defenses against pests. Even while it is growing, the more stressed a plant is, that is, the more it is subjected to pests, leaf molds and other blights, the more natural chemical pesticides it will generate."

Plant Breeders

Ames warns, "One consequence of our disproportionate concern about tiny traces of synthetic pesticide residues is that plant breeders are developing highly insect-resistant plants, thus creating other potentially more serious risks."

"For example," Ames tells us, "a major California grower introduced a new variety of highly insect-resistant celery into the market. Suddenly, the Center for Disease Control was barraged with complaints from people who handled this celery because they developed a severe rash when they were subsequently exposed to sunlight.

"Sharp detective work uncovered the fact that this pest-resistant celery had 10 times the level of psoralens, which are light-activated carcinogens, Yet this 'organic' product is still on the market."

Ames' organic foods critics argue that synthetic carcinogens are more dangerous to human health than "natural" ones,

> **BACK TO NATURE**
>
> *Abandoning modern agricultural practices, including chemical pesticides, and "going back to nature" is available to anyone who wishes to raise crops "organically" and then convince consumers that the wormy vegetables and blemished fruit are somehow worth their considerably higher price. That hardly any commercial farmers have chosen this route speaks to its impracticality. Apparently it is true that "there just are not enough people who prefer wormy apples to make them profitable. American consumers demand high quality food and remain to be convinced that they should sacrifice this quality for nebulous and nonexistent benefits of "organic" foods.*
>
> American Council on Science and Health, "Pesticides: Helpful or Harmful", September, 1988, p. 42

because the plants' natural anti-carcinogens can't deal with them.

Ames said, "That's nonsense. Humans and animals have many general defenses such as anti-oxidants that simply do not distinguish between natural or synthetic carcinogens."

Natural vs. Synthetic

What troubles Ames the most is the notion pushed by environmentalists that "natural" is "benign", while "synthetic" is automatically harmful and dangerous.

"If this were true, why are Americans healthier today than they have ever been? Why are all cancer deaths rates flat or trending down, except lung cancer, where smoking is the main cause? One reason is that thanks in part to agricultural chemistry, Americans have more fresh fruits and produce at a lower cost."

In fact, Ames said, "I usually don't buy organic food. It isn't worth the extra cost. Organic foods are selling because Americans have been sold unreasonable fears of modern technology."

Now, some in the food industry want to cash in on those fears.

READING
21
ORGANIC FARMING AND BIOTECHNOLOGY

BOTH ORGANIC FARMING AND PESTICIDES ARE NEEDED

John Hood

John Hood is a contributing editor to Reason *magazine. He wrote the following article for* The Freeman.

Points to Consider:

1. What is America's "Green Revolution"?

2. Why is the food supply safe?

3. How has the federal government hurt agriculture?

4. Why should farmers follow market forces instead of signals from Washington?

5. What is the best solution to the problems of American agriculture?

John Hood, "America Needs Organic Farming and Pesticides," *The Freeman*, February, 1990, pp. 57–58.

Growing one crop in one region may require use of synthetic pesticides and fertilizers. Farming another crop in another field might be done cheaply and productively with purely organic methods.

Organic farming is all the rage these days. After a spring of food scares and a summer boomlet of environmentalism, a report issued last September by the National Academy of Sciences (NAS) has quickly become revealed truth to legions of editorial writers, public officials, and farming mavericks. Synthetic drugs for livestock are out. Synthetic pesticides are out. Synthetic fertilizers are out. Synthesizing itself is out.

Green Revolution

And what is in? Natural fertilizers, crop rotation, careful breeding—basic farming practices making a comeback after some 50 years of neglect. The report says that such practices can be as productive, and in some cases even more so, than the standard synthetic chemical approach.

This finding has been greeted with almost hysterical glee. Many view the popularity of organic farming as the start of America's "Green Revolution". There is no question that the public's appetite is for natural foods, for foods free of dangerous chemicals and cancer-causing preservative used on apples. Supermarkets with organic food sections find themselves making a bundle off concerned shoppers (such foods are typically much more expensive than chemical-tainted wares). Grocery chains across the country have jumped on the bandwagon by issuing "pesticide-free" pledges.

Safe Food Supply

All this hype has occurred despite the efforts of many scientists to communicate a basic message to a fearful public: the food supply is safe. Agricultural chemicals, they say, pose little, if any, risk of cancer to consumers. Bruce Ames, the noted chairman of the Department of Biochemistry at Berkeley who developed the standard "Ames" test of cancer risk, estimates that the number of cases of cancer or birth defects caused by man-made pesticide residues in food or water is "close to zero."

Even so, the NAS report does make a significant point. American farmers have used synthetic chemicals and machinery reflexively, despite evidence that organic methods can in some cases be more cost-effective and maximize long-term proficiency. And the report correctly identifies the culprit: federal commodity programs which encourage over-production and punish farmers who rotate their crops.

High amounts of nitrate are a normal component of vegetables. Nitrite can be ingested directly in the diet, since it is used to cure fish, poultry and meat. Some scientists believe that human exposure to nitrosamines resulting from the dietary ingestion of nitrate and nitrite may be a factor in cancer of the esophagus and stomach.

Source: American Council on Science and Health

Federal Programs

The Federal programs, which cover about 70 percent of farm acreage, are crop-specific, paying farmers subsidies to plant the same crop year after year. If a farmer reduces his acreage, say to leave fields fallow for a year, he gets less money. If a farmer plants alfalfa or another crop designed to replenish his soil's nutrients, he gets less money. So farmers have an incentive to keep the same amount of land planted each year in the same crop, replacing crop rotation with heavy doses of synthetic fertilizers and pesticides. As the NAS report points out, the Federal subsidies insulate farmers from the true costs of the synthetic approach: soil erosion, loss of nutrients, and increased use of expensive agricultural chemicals.

James Bovard of the Competitive Enterprise Institute has identified other Federal farm programs that distort the costs and benefits of farming practices. In dry areas, Bovard writes in his book *The Farm Fiasco*, federal irrigation projects sustain crops that otherwise would be replaced with dry-weather crops more suited to soil conditions. Federal disaster payment and drought

> **ORGANIC FIELDS**
>
> *A study in* Nature *looks at neighboring farms in Oregon, both growing winter wheat. One has been organically managed since 1948, and the other has used increasing amounts of chemical fertilizers and pesticides. Today, 40 years later, the organically farmed field has six more inches of topsoil than its chemically fertilized neighbor. The organic field has a better water-storage capacity. And the topsoil on the chemical farm "will be lost in 50 years."*
>
> Tristram Coffin, Washington Spectator, *1988*

insurance reduce the financial risk to farmers who plant in low-yield or highly erosive soils, since catastrophe becomes a government problem rather than a private one. Insurance programs also relieve farmers of the need to diversify, thus placing the entire operation in greater risk when pests, weather, or disease wipe out a particular crop.

Market Forces

All these government programs are defended with the argument that the free market, which operates most of the American economy, simply does not work in agriculture. But the fact is market forces have not been allowed to impose the costs of farming methods on those who practice them. Farmers are making their decisions based on signals from Washington rather than signals from their own fields. In this way, government has established a bias toward synthetic approaches and away from organic farming, distorting agricultural markets and costing taxpayers $25 billion a year in federal subsidies.

Unfortunately, the economics behind organic farming are being overshadowed by its supposed health benefits. Instead of recognizing agricultural chemicals as tools that have been overused, many policy makers and environmental groups are treating them as poisons to be discarded. Instead of ending the current bias toward synthetic chemicals, they would replace it with a bias against them.

Banning Pesticides

Groups such as the National Toxics Campaign and National Resource Defense Council (they started the Alar scare) have long pushed for a ban on many pesticides deemed carcinogenic, and they hope the new impetus toward organic farming will increase support for such measures in Congress. Many sympathizers look to farm programs as leverage for enacting the environmentalist agenda by making "nature-conscious" practices a condition for receiving subsidy checks. Even John Pesek, the Iowa State University agronomist who headed up the NAS research effort, says that "the growing demand for safer food and a cleaner environment suggests the time is ripe" for organic farming.

But the fact is that foods grown without agricultural chemicals are rarely more safe, and sometimes are less so, than foods grown with chemicals. "All plants produce their own natural pesticides to protect themselves against fungi, insects, and predators such as man," says Ames of Berkeley. Tests of these natural pesticides have revealed that about the same percentage cause cancer in laboratory animals (30 percent) as do synthetic pesticides. Cancer-causing agents occur naturally in such foods as mushrooms, cabbage, broccoli, pineapples, and carrots.

Ames and other scientists are not saying that Americans are at high risk of cancer. Both naturally occurring and synthetic pesticides pose a negligible risk of cancer in the doses found in foods. What they are saying is that man-made chemicals are no more dangerous than those produced by the plants themselves.

In fact, breeding plants to be highly resistant to pests or disease—an approach favored by environmentalists and the NAS report to reduce the need for synthetic chemicals—doesn't always make the plants more healthful. In one case, a new variety of celery that was highly insect-resistant was introduced in California. When people handling the vegetable began to complain of severe rashes, researchers found out that it had 10 times the level of a natural carcinogen found in regular celery. "Many more such cases are likely to crop up," says Ames, "because there is a fundamental trade-off between nature's pesticides and man-made pesticides."

Nevertheless, organic farming proponents are basing their case chiefly on the specious food safety issue, when the focus should be on productivity and efficiency. Merely changing federal farm programs to encourage crop rotation instead of synthetic chemicals, on environmental or food safety grounds, would be as big a mistake as the previous policy has been.

Government Programs

Each farmer's case is different. Growing one crop in one region may require use of synthetic pesticides and fertilizers. Farming another crop in another field might be done cheaply and productively with purely organic methods. Even then, when pests or diseases suddenly strike, chemicals still may be the only effective response.

Heavy-handed government involvement, no matter how well-intended, cannot reflect these case-by-case concerns as well as can individual farmers operating in a free market. The best solution to America's farming woes is to stop treating farmers like wards of the state and start treating them like business people. Eliminate the subsidies. Let farmers decide how to plant their crops without interference from Washington bureaucrats or phobic environmentalists. Besides ending our silly bias against organic farming, we could save $25 billion from the federal budget to invest in more worthwhile pursuits, of which there is, indeed, a bumper crop.

READING

22 ORGANIC FARMING AND BIOTECHNOLOGY

BIOTECHNOLOGY AND AGRICULTURE: THE POINT

W.P. Norton

W.P. Norton wrote this article in his capacity as an editorial intern for The Progressive *magazine.*

Points to Consider:

1. What is BGH and who makes it?
2. Explain the health effects of BGH.
3. How does BGH threaten the family farm?
4. What relationship does BGH have to the farm surplus?
5. How much money have corporations spent developing and testing BGH?

W.P. Norton, "Just Say Moo: Now They Want to Drug the Cows," *The Progressive,* November, 1989. Reprinted by permission from *The Progressive,* 409 East Main Street, Madison, WI 53703.

"The time has come for the dairy industry to just say moo to crack for cows."

Milk—nature's perfect food. But will it stay that way? Not if the chemical companies have anything to do with it.

Monsanto, American Cyanamid, Eli Lilly, and Upjohn are promoting Bovine Growth Hormone (BGH), a bioengineered substance injected into cows that is said to increase milk production by as much as 20 percent.

This is the most elaborate—and expensive—attempt to move a bioengineered product from the laboratory to the marketplace. Already, consumers in some states have unknowingly been drinking this high-tech milk.

Resistance

But resistance is great at the grass roots, in the state houses, and in the boardrooms of the nation's supermarket chains, which have refused to accept dairy products from cows injected with BGH.

"BGH is a product with no redeeming value," says Jeremy Rifkin of the Foundation on Economic Trends. "It's bad for the farmer, the cow, the consumer, and the taxpayer."

The battle over BGH may be only the first skirmish in a revolution that promises to transform one of the nation's basic industries—possibly destroying a way of life in the process. Family farms may be a thing of the past as chemical companies peddle a startling range of genetically engineered food products onto supermarket shelves.

Monsanto has already sunk $1 billion into long-term biotechnology research, with most of its eggs in the BGH basket. Once in common use, the substance will generate some $500 million annually, Monsanto predicts.

With profits on the line, Monsanto has been lobbying hard. In Wisconsin alone, the corporation spent $37,833 in lobbying fees and expenses in the year starting July 1, 1988, according to the secretary of state's office. Eli Lilly and Co., based in Indianapolis, spent $20,214. Upjohn spent only $624, and American Cyanamid has no lobbyist, according to the records.

Health Effects

But the health effects of BGH milk remain in dispute. Dr. Samuel Epstein was one of the first scientists in the nation to sound the alarm on BGH. In July, barely a month before the controversy made national headlines, the University of Illinois professor of environmental and occupational medicine produced a paper claiming a wide range of harmful side-effects attributable to the consumption of BGH milk.

Reprinted with permission of *Star Tribune*.

He said substances known as "cell-stimulating growth factors" had been reported in milk from cows injected with the drug. These substances could induce premature growth and breast stimulation in infants and possibly promote breast cancer, Epstein warned. There was, he believed, enough uncertainty over the behavior of genetically engineered BGH to warrant the start of long-term human health studies.

The U.S. Food and Drug Administration, he claimed, "acted irresponsibly and recklessly" in approving the drug for testing and consumption, and "abrogated its regulatory responsibilities in favor of protecting industry interests."

It was a matter, Epstein said, of the public being kept in the dark by companies that had reason to fear their massive investments were being jeopardized.

"In the majority of instances where industry interests are involved, the only information available to Congress and decision makers is either developed in-house, or by contract with university research arms," he said. "This creates an inherent conflict of interest, and a conscious and subconscious effect on those receiving the funds."

But Dr. Henry A. Miller, the special assistant to the FDA Commissioner on biotechnology issues, ridiculed the safety issue, calling Epstein's report a "gross distortion of scientific

> **MILK IS UNSAFE WHEN COWS ARE JUNKIES**
>
> *America's dairy cows—among the most overworked and exploited animals in the nation—are shot through with so many drugs that milk ought to be sold as a prescription-only product. Some drugs are antibiotics to prevent disease, others are tranquilizers to ease dairy factory tensions. Some are booster-shot hormones that double or triple milk production, creating super-cows that have udders of such size and low-slung weight that walking requires exhaustive effort. At many dairies, cows don't walk. They are penned in storage bins and, twice daily, transported on flat escalators to and from the milk parlor.*
>
> Colman McCarthy, Washington Post, *February, 1990*

fact."

Threat to Consumers

The extent of any threat to consumers remains unclear. The longest-term studies of BGH on record so far do not stretch over the five-year milking life of a typical dairy cow. Besides the Epstein report, virtually nothing has been said to contradict the FDA's assertion that the drug cannot be detected in milk and would have no secondary effects on humans.

The agency does have BGH studies conducted by the chemical giants, but has declined to make them public. "It's outrageous that FDA refuses to release these studies," says Wisconsin Secretary of State Doug LaFollette. "I've been trying to get them for weeks."

But syndicated columnist Jack Anderson says he obtained copies of the confidential documents showing some cows that received BGH injections "lost weight, suffered lower fertility rates or anemia, or came down with mastitis (inflammation of the mammary glands)." In some cases, Anderson wrote, Monsanto researchers refused to count cows that got mastitis.

Concerned about the health effects, Rifkin's foundation blitzed the nation's twelve largest supermarket chains with letters outlining the safety concerns and underscoring consumer polls indicating a severe backlash against milk products containing BGH.

Meanwhile, the chains' own consultants were citing a 1987 report from the Office of Technology Assessment which found that 52 percent of Americans believed genetically engineered food products may pose a serious danger to people or the environment. Market analyses also showed consumers are repelled by the mere mention of hormone additives in milk. . .

Family Farm

BGH and products like it could eventually make it possible for only 50,000 commercial farms, one-tenth of the present number, to produce most of the nation's food. Smaller farms and the communities that support them would be the first casualties.

The family farmer is already an endangered species. Small farming is only now picking itself up from the farm crisis of the mid-1980s, which drove thousands out of a time-honored way of life.

One who stayed on the land is dairy farmer John Kinsman, whose family has tilled the soil of Lime Ridge, Wisconsin, for generations. Kinsman has been in the forefront of opposition to BGH.

"All we've ever asked is to give the public both sides and let them decide," he says. "Now, after four years, we are more convinced than ever of the value of working to stop this. BGH is the tip of the iceberg of what these oppressive industries and their allies want to put on the consumers of the world."

With 17 percent of the nation's milk production, Wisconsin is the top dairy state. So it seemed fitting that the fight over BGH would be joined there. . .

Kinsman sees the fight against BGH and the fight to save the family farm as one and the same. "It's more than just BGH that we're opposing," he says. "This product and what it represents probably pose far more serious hazards than did the nuclear revolution. Consumers in large numbers are beginning to understand how important it is that the family farm is preserved, because once that is gone, there is no watchdog for the quality of purity of the food they get."

BGH may not be good for cows, either. "It's just going to create another drug dependence," says Dr. Michael Fox, a veterinarian and head of the National Humane Society. "High-yielding dairy cows already suffer from a wide spectrum of diseases. The cow's immune system will be placed in jeopardy because of energy budgeting: she's putting so much into milk output that her own bodily needs will be sacrificed. We'll see a faster turnover rate, and more cows going into hamburgers. . ."

One of their tactics is to avoid any reference to genetic engineering, according to an analyst quoted in *The Wall Street Journal.* That's why the industry insists on using the more scientific — and sanitized — term Bovine Somatotropin, or BST, rather than BGH.

The FDA is holding steadfast to its position that BGH milk is safe. It's not the first time the agency has insisted that a radical new chemical is benign: such pesticides as DDT, synthetic es, and chemical additives now known for their high risks once touted as threat-free.

Surplus

the moment, however, the health issue is overshadowed by farmers' concern over the all but certain concentration of ownership that will affect American agriculture after BGH.

Since 1979, the Government has spent almost $16 billion to buy and store surplus butter, milk, and cheese. The nation was so glutted with milk that by 1986 and 1987, the Government picked up the tab for the most drastic surplus-cutting measure yet — a $1.8 billion program to pay 14,000 farmers for slaughtering 1.5 million cows and calves. The price tag on the dairy surplus program was about $12 billion.

With so much milk around, and production showing no signs of slowing down, Wisconsin farmers alone would lose as much as $100 million annually if BGH takes off. The country's 160,000 dairy farmers produce about 150 billion pounds of milk a year. If, as predicted, BGH drives down milk prices, it would eventually squeeze thousands of farmers off the land, leaving only the corporate farms with large-scale managerial resources and deep pockets to weather the storm.

BGH combatants are already drawing the battle line for the next round. "We're fighting on two levels," says John Stauber, Jeremy Rifkin's associate in Wisconsin. "Our biggest success is going to come in the nonlegislative arena. If we're successful there, the rest will follow."

Opposition

In every major dairy state from Vermont to California, activists are organizing an insurgency that draws on the long-dormant populist roots of American agriculture. They are organizing dairy farmers into coalitions which, in turn, demand that their dairy co-ops go on record as refusing to take milk from BGH cows.

If the cooperative says no, the farmers are encouraged to sell their milk elsewhere. "It's a very powerful economic incentive," says Stauber. "We want to make it so there won't be any processor in the country who will handle this milk."

Still, farmers appear to be hedging their bets. While 75 percent of dairy producers oppose FDA approval of the hormone, only 28 percent say they would not use it, according to a national survey by the publication *Dairy Herd Management*.

And testing involving some 11,000 cows around the country goes on. Field trials and university studies are being conducted in at least twenty states at a cost of $10 to $15 million a year.

Approval of BGH by the FDA for commercial sale is a near certainty within a few months. The only delay would come from

Washington, where Congress could call for further studies on the matter, or even a commercial moratorium.

The Corporations

But in the long run, time and money are on the side of the corporations. "Companies like Eli Lilly and Upjohn will clean up on this stuff if it gets approved," Stauber says. "If the cows get sick, they'll just make more money selling antibiotics to the farmers. They just can't lose."

Biotechnology may be here to stay, but it's up to the public to decide whether we have anything to say about it. Without a sustained public outcry, the companies and their friends in the Government regulatory agencies will continue to tinker with life's most basic processes.

"The time has come for the dairy industry to just say moo to crack for cows," says Michael Picker of the California Biotechnology Action Council, which opposes BGH. "But these guys are really hooked on all these pharmaceutical inputs. It's hard for them to resist. It's a real high for the corporate dairy industry."

READING

23 ORGANIC FARMING AND BIOTECHNOLOGY

BIOTECHNOLOGY AND AGRICULTURE: THE COUNTERPOINT

John Nicholson

Mr. Nicholson, a Washington free-lance writer, operates a public affairs firm that has done work in the past with several of the major chemical companies producing the new BST. (BST) Bovine Somatotropin is the more scientific name for (BGH) Bovine Growth Hormone.

Points to Consider:

1. What warning did Congressman Steve Gunderson give to Wisconsin farmers?

2. Explain the nature and purpose of the new product called Bovine Somatotropin (BST).

3. How have consumers reacted to BST?

4. Why is BST safe to use?

5. Who is Benny Haerlin and what did he do?

John Nicholson, "Germany's Green Fuels U.S. Farm Controversy," *Human Events*, April 1, 1989, p. 11

BST will be the first popular use of biotechnology related to the retail marketing of milk.

International leaders of the radical Germany-based Green Party are aiming at U.S. state legislative arenas such as Wisconsin to establish beachheads, much in the same way they've discovered how to work quietly on the local level before winning successes in the European Community's overall Parliament.

Wisconsin state politicians like State Senator Russ Feingold deny being exploited by Green party advocates in a still-developing battle over the use of new technologies in dairy production.

Rep. Steve Gunderson (R—Wis.) warned the Wisconsin Farmers Union that efforts to ban the new dairy product known as Bovine Somatotropin will only isolate the state's milk producers; he said Wisconsin doesn't "have the luxury to answer the BST question by itself. This is a national and international question."

New Product

At issue is the use of a new product soon to be approved by the FDA, called "bovine somatotropin," or BST. The hormone now exists in milk in trace amounts and is digested easily by humans as a harmless protein.

FDA has taken much heat in approving meat and milk for human consumption resulting from the field trials of BST supplements, but no one disputes its safety from a human consumption standpoint.

As Senator Feingold alleges, "Consumers apparently don't want to eat and drink dairy products that have been treated with this hormone—even though no ill effects are known." (Industry spokesmen point out that products are not "treated" with BST, but the facts don't deter the politicians' labels.)

> Instead, the activists claim the use of BST will overstress cows, an animal welfare topic, and many small-herd dairy farmers fear the larger "factory" operations will use the new technology to exacerbate the slow demise of the less efficient family farm.

Produced by the pituitary gland, BST regulates the milk production system of the cow; for decades farm scientists have known that more BST makes cows more productive. Not until now, however, has there been a way to reproduce the BST economically—thanks to biotechnology—so dairy farmers can get more milk from the same herd by supplementing the natural BST. And tests show there are no higher BST levels in milk from cows receiving the supplement.

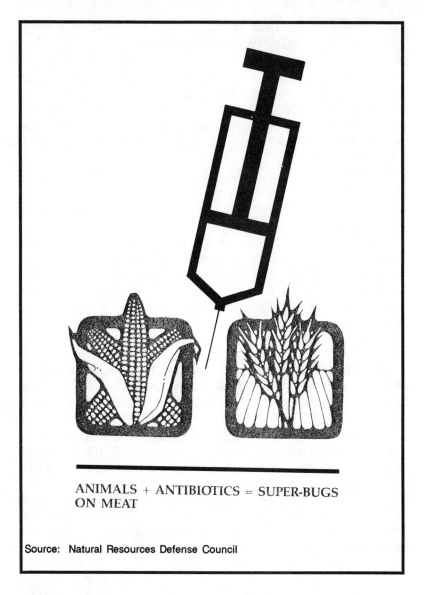

ANIMALS + ANTIBIOTICS = SUPER-BUGS ON MEAT

Source: Natural Resources Defense Council

Biotechnology

BST will be the first popular use of biotechnology related to the retail marketing of milk. It sounds synthetic, although it's duplicated by natural means under controlled bio-tech circumstances. BST benefit is only economic, and initial trials indicate it will cause 15 to 25 percent increases in production, which some family farmers think is hardly worth the effort with a small herd. Producers point out, however, that use of BST is one way to respond to price fluctuations by augmenting milk flows (it works virtually overnight) without having to add to their herd.

> **DON'T BAN HORMONE**
>
> Minnesota's agriculture commissioner said Friday would be a mistake for the state to prevent farmer using a genetically engineered hormone that would milk production in cows.
>
> Commissioner Jim Nichols said he doesn't like the idea of using the hormone, but believes Minnesota's dairy farmers would be at a disadvantage if the state enacts a ban when other states don't.
>
> Jim Parsons and Sharon Schmickle, Star Tribune, April 1, 1989, p. 3B

Dr. Richard Phipps, a scientist at the government-supported Institute of Grassland and Animal Production at Shinfield, England, recently refuted at a conference the so-called "instant experts" who allege that BST will lead to metabolic breakdown, cows burning themselves out after two or three lactations, and a high incidence of milk fever ketosis. "Over 150 papers have been reviewed in competent scientific journals and not one of them has referred to these metabolic problems," he said. "Indeed, in our own long-term study we have not recorded one single case of ketosis, so I think you can forget about this."

Small Herds vs. Large Herds

At that conference, a leading U.K. cattle specialist, Allan Buckwell of Wye College, pinpointed the economic division between small-herd family operators and large-scale dairy operations. "In my view," he said, "the only people who can profitably use BST on a programmed basis, on their whole herd just about all of the time, are those who know how to get high yields using few concentrates—in other words, the efficient feeders of cows."

German Green party leader Benny Haerlin skillfully interweaves the animal rights issues with the economic cross-fire and the general fear of something new into a seductive, anti-corporate, anti-big business pitch. He kept the Wisconsin legislative hearing spellbound last fall by asserting that the advent of BST "will not simply go away; it has to be defeated politically." And he claimed that in every country "threatened" by the four (American Cyanamid, Monsanto, Lilly and Upjohn) BST-producing U.S. companies, "a remarkable new coalition is forming." . .

"For the first time groups often in conflict are working together. Farmers, consumers, animal welfarists, religious ethicists and environmentalists are uniting. This effort is growing

...rldwide, as it must to protect our family farms, and to protect the purity and integrity of our dairy products."

Political Agitators

The international network of political agitators is watching this case history closely to see if the mystique of a spokesman from some other land can be used successfully in American politics. Within the European Community, agitators have discovered a fuss in one country can be imposed on other nations by adroit maneuvering in the European Parliament, especially by waving the flag against big business and raising as many alarming ramifications as possible.

Haerlin is a co-founder of the Genethischez Netzwerk (Gen-ethic Network), set up in 1986 to stop BST in Europe. Closely aligned with the Green Movement, the Network sets as a goal to "empower...citizens to more effectively participate in the development of public policy to guide genetic engineering technology." It has a working relationship with the Boston-based Committee for Responsible Genetics, as well as the Feminist International Network of Resistance to Reproductive and Genetic Engineering (FINRRAGE), to orchestrate the groups' attacks...

Other Issues

Using the BST controversy to raise other issues is a Green party tactic that has attracted Doug LaFollette, Wisconsin's Secretary of State, who holds a Ph.D. in organic chemistry and is a major environmentalist. He opposes BST because, he says, he's nervous "when scientists who should know better fail to ask the right questions and tell us there's no problem with a new, untested, unnecessary technology." Furthermore, he asks:

"What will happen to our rural communities if the dairy farms surrounding them are forced out of business, not only in Wisconsin but in Vermont, Iowa and Minnesota as well? We are faced with the prospect of having a few giant corporate farms with cows used as 'milk machines' and ghost towns everywhere else."

At the center of this fuss stands Jeremy Rifkin, a short, bald, 43-year old self-professed heretic described by the *New York Times* as pursuing causes "like the 19th Century machine-bashing Luddites." Rifkin "has poisoned the whole atmosphere around which biotechnology has developed, rather than allowing it to be developed in a rational and thoughtful manner," says Noble laureate David Baltimore from the Massachusetts Institute of Technology...

European Agenda

Whether or not Senator Feingold and others will remain duped by the axis of Rifkin and the Green party is still to be played out. Last month's vote in the state's farm bureau convention suggests farmers, at least, are more swayed by the potential problems through biotechnology. So long as local politicians can derive enthusiasm (or other support) from the Green party, the international agitators appear to have found some willing partners at the local levels of U.S. politics, ready to exploit a new and as yet undetermined radical European agenda.

Reading and Reasoning

WHAT IS EDITORIAL BIAS?

This activity may be used as an individualized study guide for students in libraries and resource centers or as a discussion catalyst in small group and classroom discussions.

The capacity to recognize an author's point of view is an essential reading skill. The skill to read with insight and understanding involves the ability to detect different kinds of opinions or bias. Sex bias, race bias, ethnocentric bias, political bias and religious bias are five basic kinds of opinions expressed in editorials and all literature that attempts to persuade. They are briefly defined in the glossary below.

Glossary of Terms for Reading Skills

SEX BIAS — the expression of dislike for and\or feeling of superiority over the opposite sex or a particular sexual minority.

RACE BIAS — the expression of dislike for and\or feeling of superiority over a racial group.

ETHNOCENTRIC BIAS — the expression of a belief that one's own group, race, religion, culture or nation is superior. Ethnocentric persons judge others by their own standards and values.

POLITICAL BIAS — the expression of political opinions and attitudes about domestic or foreign affairs.

RELIGIOUS BIAS — the expression of a religious belief or attitude.

Guidelines

1. From the readings in Chapter Five, locate five sentences that provide examples of editorial opinion or bias.

2. Write down each of the above sentences and determine what kind of bias each sentence represents. Is it *sex bias, race bias, ethnocentric bias, political bias, or religious bias?*

3. Make up one sentence statements that would be an example of each of the following: *sex bias, race bias, ethnocentric bias, political bias and religious bias.*

4. See if you can locate five sentences that are factual statements from the readings in Chapter Five.

BIBLIOGRAPHY

Allera, Edward J. An overview of how the FDA regulates carcinogens under the Federal Food, Drug, and Cosmetic Act. Food Drug Cosmetic Law Journal, v. 2, Feb. 1978: 59-77.

American Meat Institute. Internal document stating reasons in opposition to labeling imported meats, May 1985.

Ames, B. N. (1983). Dietary Carcinogens and Anti-carcinogens. Science 221 1256-1264.

Ames, B. N., Magaw, R., Gold. L. S. (1987). Ranking Possible Carcinogenic Hazards. Science 236, 271-280.

Burbee, Clark R., and Carol s. Kramer. Food Safety Issues for the Eighties. National Food Review, spring 1986: 17-20.

Council for Agricultural Science and Technology. Regulation of Potential Carcinogens in the Food Supply: The Delaney Clause. Ames, Iowa, 1981. 15 p.

Commoner, B., Vithayathil, A. J. Dolara, P., Nair, S., Madyastha, P. and Cuca, G. C. (1979). Formation of Mutagens in Beef and Beef Extract During Cooking. Science 201:913-916.

Demkovich, Linda E. Critics fear the FDA is going too far in cutting industry's regulatory load. National Journal, v. 14, July 17, 1982: 1249-52.

Doll, R. and Peto, R. (1981). The Causes of Cancer: Quantitative Estimates of Avoidable Risks of Cancer in the United States Today. JNCI 66: no. 6, 1193-1308.

Falk, R. T., Pickle, L. W., Fontham, E. T., Correa, P., and Fraumeni, J. F. Jr. (1988). Life-style Risk Factors of Pancreatic Cancer in Louisiana: A Case-control Study. American Journal of Epidemiol.:128 no. 2, 324-336.

FDA 75 years. FDA Consumer, v. 15, June 1981: whole issue.

Food Chemical News. Washington, Food Chemical News Inc. Weekly. 1958 to present.

Food Safety Council. A proposed food safety evaluation process: Final Report. Washington, 1982. 142 p.

Gilbert, Susan. America Tackles the Pesticide Crisis. The New York Times Good Health Magazine, Oct. 8, 1989: 22-25, 51-52,

54-57.

Goyan, Jere E. The future of the FDA under a new administration. Food Drug Cosmetic Law Journal, v. 36, Feb. 1981: 60-65.

Hall, Richard L. Food Safety and Risk/Benefit. Research Management, v. 24, Jan. 1981: 28-35.

Hatch, F. T., Felton, J. S., Stuermer, D. H. and Bjeldanes, L. F. (1984). Identification of Mutagens from the Cooking of Food. In "Chemical Mutagens, Principles and Methods for Their Detection," (Ed. de Serres, F. J.) Plenum Press, New York, pp. 111-164.

Havender, William. The Science and Politics of Cyclamates. Public Interest, v. 71, spring 1983: 17-32.

Investigating Food Safety. National Food Review, v. 12, July-Sept. 1988: whole issue (39p.).

Litman, Richard and Donald Litman. Protection of the American consumer: The Congressional battle for the enactment of the first federal food and drug law in the United States. Food Drug Cosmetic Law Journal, v. 37, 1982: 310-329.

Millstone, Erik. Food Additives: A Technology Out of Control? New Scientist, v. 104, Oct. 18, 1984: 20-24.

Morrison, R. M., and T. Roberts. Food Irradiation: New Perspectives on a Controversial Technology. U.S. Department of Agriculture. Economic Research Service. Dec. 1985.

Mott, L. and Snyder, K. (1988). "Pesticide Alert" The Amicus Journal Spring NRDC, Publications Department, 122 E. 42nd Street, New York, N.Y., pp. 20-29.

National Fisheries Institute. U.S. Seafood Processing Industry: An Economic Profile for Policy and Regulatory Analysts. Study prepared for NMFS, Washington, D.C., 1983. p. 27.

NAS (1981). Committee on Nitrite and Alternative Curing Agents in Food, Assembly of Life Sciences, National Academy of Sciences, "The Health Effect of Nitrate and N-Nitroso Compounds," National Academy Press, Washington, D.C.

National Research Council. Committee for a Study of Saccharin and Food Safety Policy. Saccharin: Technical Assessment of Risks and Benefits. Part 1 of study. Washington, National Academy Press, 1978, 1 v.

National Research Council. Food safety policy: Scientific and Societal Considerations, part 2 of study. Washington, National Academy Press, 1979, 1 v.

National Research Council (1982). "Report on Diet, Nutrition and Cancer" (National Academy of Sciences, Washington, D.C. pp. 12-6.

National Research Council (1987). Regulating Pesticides in Food, The Delaney Paradox Committee on Scientific and Regulatory Issues Underlying Pesticide Use Patterns and Agricultural Innovation, Board of Agriculture, Washington, D.C., pp. 45-99.

National Research Council (U.S.) Board on Agriculture. Committee on Scientific and Regulatory Issues Underlying Pesticide Use Patterns and Agricultural Innovations. Regulating Pesticides in Food: The Delaney Paradox. Washington, National Academy Press, 1987. 272 p.

NBS Develops Method for Detecting Food Irradiation. Food Chemical News, June 23, 1986.

Neff, D. The Regulation of Food Irradiation. Nuclear Law Bulletin, no. 35, June 1985.

Newberne, P.M. and Suphakarn, V. (1983). Nutrition and Cancer: A Review, with Emphasis on the Role of Vitamins C and E and Selenium. Nutrition and Cancer 5(2):107-119.

Perera, F. and Boffetta, P. (1988). Perspectives on Comparing Risks of Environmental Carcinogens. Journal of the National Cancer Institute 80: 1282-1293.

Public Voice for Food and Health Policy. A Market Basket of Food Hazards: Critical Gaps in Government Protection. [by] Thomas Smith. Washington, April 1983. 106 p.

Reed, Donald, Pasquale Lombardo, John Wessel, Jerry A. Burke and Bernadette McMahon. The FDA Pesticide Monitoring Program (Unpublished draft, available through the Division of Chemical Technology, Food and Drug Administration).

Reed, Donald. The FDA Surveillance Index for Pesticides: Establishing Food Monitoring Priorities Based on Potential Health risk. Journal of the Association of Official Analytical Chemists, v. 68 (1982).

Smith, Thomas B. A Market Basket of Food Hazards: Critical Gaps in Government Protection. Washington, Public Voice for Food and Health Policy, 1983. pp. 10-19.

Stoloff, L. (1977). Aflatoxins-An Overview, In "Aflotoxins in Human and Animal Health" (Eds. Rodricks, J. V., Hesseltine, C. W. and Mehlman, M. A.) pp. 7-28. Pathotox Publishers, Inc., Park Forest South, Illinois.

Sugarman, Carole. Who's Minding the Store? Washington Post, Oct. 4, 1989: E1, E5.

Thompson, Richard. Purifying Food Via Irradiation, FDA Consumer, Oct. 1981.

United Nations. Food and Agriculture Organization. Wholesomeness of Irradiated Food. Report of the Joint FAO/IAEA/WHO Expert Committee, no. 6. Rome, 1977.

Urbain, W. M. Irradiated Foods: A Giant Step Beyond Appert. Nutrition Today, July/Aug. 1984.

U.S. Congress. House. Committee on Agriculture. Subcommittee on Department Operations, Research and Foreign Agriculture. Hearings, 97th Congress, 1st session, on H.R. 638–National Science Council. Washington, U.S. Govt. Print. Off., 1981. 312 p.

U.S. Congress. House. Committee on Agriculture. Subcommittee on Livestock and Grains. Review of FSQS operations. Hearings 95th Congress, 2d session. July 11, 1978. Washington, U.S. Govt. Print. Off., 1978. 124 p. "Serial No. 95-JJJ"

U.S. Congress. House. Committee on Government Operations. HHS' failure to enforce the Food, Drug, and Cosmetic Act: The Case of Cancer-Causing Color Additives. Washington, U.S. Govt. Print. Off., 1985. 73 p. (99th Congress, 1st session. House. Report no. 99-151)

– – –The regulation by the Department of Health and Human Services of carcinogenic color additives. Hearing, 98th Congress, 2nd session. Oct. 5, 1984. Washington, U.S. Govt. Print. Off., 1977. 738 p.

U.S. Congress. House. Committee on Government Operations. Subcommittee on Intergovernmental Relations and Human Resources. Regulation of carcinogenic additives. Hearing, 100th Congress, 1st session. June 24, 1987. [unpublished hearing record]

U.S. Congress. Committee on Interstate and Foreign Commerce. Subcommittee on Health and the Environment. Saccharin ban—National Academy of Sciences on the use of saccharin. April 11, 1979. Washington, U.S. Govt. Print. Off., 1979. 48 p.

U.S. Congress. Senate. Committee on Agriculture, Nutrition and Forestry. Food safety: Where are we? Washington, 1979. 678 p. At head of title: 96th Congress, 1st session. Committee print.

U.S. Congress. Senate. Committee on Human Resources. Subcommittee on Health and Scientific Research. The banning of saccharin, 1977. Hearings, 95th Congress, 1st session, on examination of the risks included in the use of saccharin and the decision by the Food and Drug Administration to ban the substance from the market. June 7, 1977. Washington, U.S. Govt. Print. Off., 1977. 173 p.

U.S. Congress. Senate. Committee on Labor and Human Resources. Oversight on food safety, 1983. Hearings, 98th Congress, 1st session. June 8-10, 1983. Washington, U.S. Govt. Print. Off., 1983, 666 p.

U.S. Congress. Senate. Committee on Labor and Human Resources. S. 484, a bill to extend the moratorium on the ban of saccharin for three years. Hearing, 99th Congress, 1st session. April 2, 1985. Washington, U.S. Govt. Print. Off., 1985.

U.S. Congress. Senate. Select Committee on Small business. Food additives: Competitive, regulatory and safety problems, part I and II. Hearings, 95th congress, 1st session. January 13-14, 1977. Washington, U.S. Govt. Print. Off., 1977. 979 p.

U.S. Department of Agriculture. Food Safety and Inspection Service. Meat and Poultry Inspection, 1984: Report of the Secretary of Agriculture to the Congress. March 1985. Washington, 1985. p. 27.

U.S. Department of Agriculture. Food Safety Inspection Service. Letter to House Agriculture Committee in response to request for comment on H./R. 3670, June 1981.

U.S. Department of Agriculture. Economic Research Service. Economic Effects of Requiring Imported Meats To Be So Labeled. 1978.